Studies in Computational Intelligence 423

Editor-in-Chief

Prof. Janusz Kacprzyk
Systems Research Institute
Polish Academy of Sciences
ul. Newelska 6
01-447 Warsaw
Poland
E-mail: kacprzyk@ibspan.waw.pl

For further volumes:
http://www.springer.com/series/7092

Yukio Ohsawa and Akinori Abe (Eds.)

Advances in Chance Discovery

Extended Selection from International Workshops

Springer

Editors
Prof. Dr. Yukio Ohsawa
The University of Tokyo
Bunkyo-ku
Japan

Akinori Abe
Faculty of Letters
Chiba University
Chiba
Japan

ISSN 1860-949X e-ISSN 1860-9503
ISBN 978-3-642-30113-1 e-ISBN 978-3-642-30114-8
DOI 10.1007/978-3-642-30114-8
Springer Heidelberg New York Dordrecht London

Library of Congress Control Number: 2012937234

Printed on acid-free paper

Springer is part of Springer Science+Business Media (www.springer.com)

Preface

Yukio Ohsawa and Akinori Abe

The beginning of year 2000, we started to study methods for discovering events/situations that significantly affect decision making. The triggers to this movement and its aftershocks occurred with the initiator Ohsawa's movements from/to schools in engineering to business sciences since around in April 1999. When he was studying methods for visualizing data on earthquakes with his original KeyGraph® (See chapters in [Ohsawa and McBurney, 2003] for some use cases), where high-risk faults of earthquakes were shown by red nodes representing rarely quaking faults affected by frequently quaking active faults visualized by black nodes, the business students who work at daytime who learned rare but significant events may get highlighted by KeyGraph suggested such a graph may be useful for aiding decision making in business.

As a result, they started to collaborate with companies to visualize data on businesses, to detect noteworthy events which might be rare but useful for planning new actions and decisions. Such events or situations are sometimes called opportunities, and other times crisis. With realizing this kind of successes (also see Fig. 1, they started series of workshops and special sessions on what they called Chance Discovery (IECON2000, KES 2000, AAAI2001, ECAI2004, etc) — second editor Abe joined these activities, and now is leading the world-wide network of researchers on chance discovery. Through these active efforts, the participants' vague dream converged to a unique definition of a chance, i.e., an event which has a significant impact on human's decision making. Thus a chance provides an uncertain opportunity/risk for human. And, chance discovery is the discovery *of* a chance, emphasized in contrast to discovery *by* chance. Because the essential aspect of a chance is that it can be the seed of new and significant benefits/loss for human, the sense of human(s) about benefit/loss is a significant factor for chance discovery as well as machine's power to analyze and visualizing data for aiding humans' talent.

Fig. 1. A successful case in the classical chance discovery: The clusters of frequently ordered items are bridged via a rare item, from which textile marketers concluded to propose a new selling scenario: Make a jacket for business people with the new corduroy, to which they can change from suits to go out for dinner after working time [Ohsawa and Usui, 2005].

As well as workshops on chance discovery, they published books and issues from journals, characterizing chance discovery with studies on:

(1) Human factors in cognitions, communications, and thoughts for catching up with a chance and understanding its significance for decision making,
(2) Process for externalizing candidates of chances and discovering chances on objective evidences — reinforcing (1) by data in other words, and
(3) Method and tools mining/visualizing data that aid in chance discovery — accelerating and sustaining the process of (2) in other words.

In other words, the design of interaction among human(s), computers, and their environment came to the core interests in studies for realizing chance discovery. The papers contributed to chance discovery so far have been by researchers from various domains such as artificial intelligence, creativity support, economics, business administration, risk management, operation research, linguistics, mathematics, physics, psychology etc showing approaches relevant to classes (1), (2), and (3).

For the present book, we selected authors of outstanding papers which were presented on chance discovery in relevant workshops, conferences, and symposia — mainly from workshops in the International Joint Conference on Artificial Intelligence (IJCAI 2011), IEEE International Conference on Data Mining (ICDM 2011), IEEE International Conference on Systems, Man, and Cybernetics

(SMC2010). Collaborators who utilized or extended studies in chance discovery for contributing to other but relevant conferences, such as of multi-agents and juristic communications (AAMAS2011, JURISN2011) were also invited. As guest editors, we are happy to find three chapters for class (1), five for (2), and seven for (3) in this book. Readers will find new visions for the decision making of human, from this harmonious combination of contributions.

In the first part, we show contributions to the aspect of Cognition and Communication toward Chance Discovery. Abe's introduction of the concept *curation* in Chapter 1 encourages us to discuss the meaning of chance discovery from a different aspect coming from the exhibition of artwork in a gallery. Borrowing ideas from cases of data curation and curation of artwork, he suggests a direction to curation for aiding chance discovery, and leads to the concept to the role of communication in chance discovery. In Chapter 2, by Bardone and Magnani, the bullshitting phenomena that is a significant kind of chance faking, is highlighted and explained with respect to one's carelessness about frames, evidences, and simplicity that one should stay aware of for taking real advantage of noteworthy chances. The messages suggested in this paper are instructive and useful for people who seek chances — we learn from this paper that we need evidential data and also concrete interests in explaining the data in the form of a simple logic under a suitable context, whether or not data mining tools are provided. Terai and Miwa links chance discovery to their original studies on *insight* in Chapter 3, that means to release humans from impasse, so that noticing the value of a new event gets enabled. They argue that in insight problem solving tasks, a prepared mind precedes the "aha" experience, as in the process to chance discovery. In order to preparing mind for chance discoveries, they point out interdisciplinary studies between cognitive researches and techniques that encourage chance discovering. Also introducing an original study attempting to reveal the prepared process of insight problem solving using eye movement data, they point out possible approaches from cognitive science to chance discovery.

All in all, the first part focuses on how we can open human's mind out to the value of external information. We have been endeavoring to reinforce such cognitive forces toward chance discovery, by developing technical aspects of human-machine and human-human interactions with developing techniques of data visualization — as in the case KeyGraph has been used for discovery of an opportunistic item in the market as shown in Fig. 1. In other words, data visualization plays the role as the environment for chance curation. That is, how to present data on facts as evidences for viewers to notice chances — this is the concept Chapter 1 introduced in this book and the second part embodies by presenting five contributions. And, this is the reason why we give the title "Data Visualization as Chance Curation" to the second part.

The second part starts from Chapter 4, where Sarlin writes an approach to chance discovery in finance, with self-organizing maps for discovering imbalances in financial networks. Here we find the diffusion of financial crisis from/to countries and financial state transitions are easy to see on the visualized maps, so that early signs of changes shall be caught as chances of people in businesses including investments and the government. In Chapter 5, Sunayama et al shows a recursive method for clustering and visualizing textual data, so that human can detect useful information from the atlas of documents. As chance discovery favors utility sometimes more than accuracy, their method which does not always present accurate clusters but provide useful information is expected to be a promising tool for chance discovery.

Another feature of the second part is that communication is introduced as a building block of chance discovery. In Chapter 6, Nitta presents a method for analyzing discussions, coupling logical analysis of the discussion subject and word statistical analysis — a temporal word clustering method — of the log text of discussion. We can appreciate this work as a way to integrate chance discovery and logical modeling of the dynamics in the target domain. By integrating these two methods, the discussion moderation skills are analyzed to find reusable knowledge for other discussion. Furthermore, nonverbal information can be targeted by extending Nitta's method. In Chapter 7, Lin et al. visualizes graphs representing the social network linked to innovation and its diffusion. By thus visualized evidences, we can understand the dynamics of innovation and detect essential events which, and early adopters who, may have caused or may cause the growth of industry. This approach is succeeding the essential feature of KeyGraph (Fig. 1), but their visualization will be found easier to see in interpret. Wang and Ohsawa, in Chapter 8, presents iChance a Web-based environment for innovative communication with chance curation. Here, the gaming principle of Innovators Marketplace® (this is Ohsawa's trademark in US: Innovators Market Game® is a trademark in Japan) has been implemented on the Web and reinforced by Wang's style of communication where inventors evaluate consumers (as well as vice versa) and his new visualization technique Gallaxy. We are showing how chance discovery can be really realized by the collective intelligence of humans and machines.

Let us look at the first and the second parts above from another aspect: While humans perceive and understand chances and plan scenarios of actions and events in the future, how are data processed? As in Fig. 2, we have been modeling the process of chance discovery as a double spiral — humans' deepening of interests in chances and data processing (collection and visualization according to humans) interest.

We can say the first part of this book has been dedicated to the humans' spiral, where collaborators communicate with curating chances, reframing, and reaching insights to revise interests to adopt to the changing real world. And, the second part extends our view to include technical aspects where machine spiral are involved. Here we notice — the viewpoints of human(s) are revised in stepping up the spiral so that the data in hand and collected from external world are viewed and interpreted in a new direction. That is, in the case of Fig. 1, two clusters of textile items, which had been respectively regarded as materials for casual clothes and for business suits, came to be interpreted as one meta-cluster created by combining the two existing clusters, by introducing a new viewpoint to satisfy business people to take easy clothes after their working time. In short,

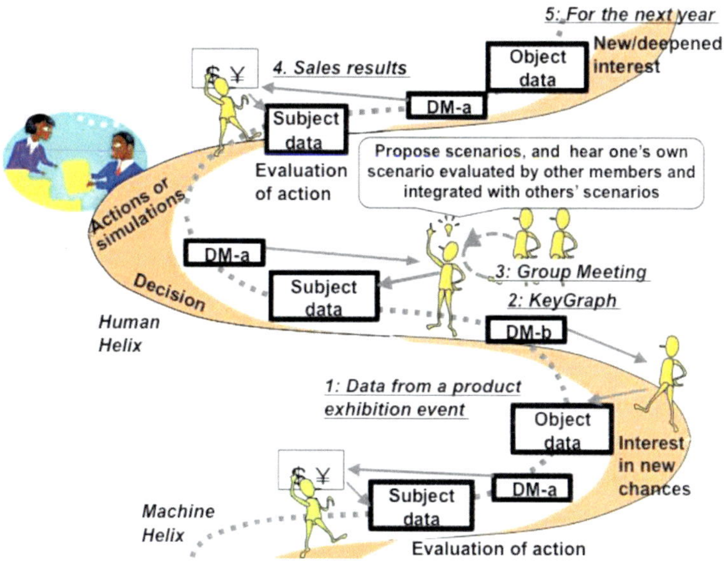

Fig. 2. Double Helix (DH) process of chance discovery (also see [Ohsawa and McBurney, 2003] — not only chapter 1 but others are directly or indirectly based on this process).

the thing which occurred here and is expected by following the DH process is the synthesis of evidences, via the analysis and visualization of data.

The third part of this book mainly highlights techniques for data mining, that may seem to essentially focus on the analytical aspect of data processing. However, we position these work as computational and logical cutting edges for not only data analysis but also data synthesis. In Chapter 9, Furuhata, Mizuta, and So shows a method for tracking concept drift, for forecasting different types of sudden unexpected changes and to be adaptive to these changes. Their forecasting method based on the stable evaluator and the reactive evaluator are good at dealing with consecutive concept drifts. An application of this method is Finance, as tested in this paper to financial data in US including the late-2000s recessions. In Chapter 10, Zhang, Leung, Pang, and Tang proposes a multi-agent system for finding satisfactory services. As this work is dedicated to the general problem to match increasing providers and recipients efficiently, reader in any service domain is expected to project this work to his/her own field e.g., how can all patients find the best doctor for the health problem of each? In Chapter 11, Pogorelc and Gams focuses join health problem: Their technique enables to discover clues for medical decision making — do nothing or call for help — from data on daily movements of elderly people. Although their approach once pulls us back to the traditional view point of data mining, that is classification accuracy, we notice there are domains where accuracy plays a critical role in quick and suitable decision making. It is thus important to consider the use scenario where data are finally synthesized into decision making, in order to choose the viewpoint in data analysis.

Thoughts about computation techniques sometimes give birth to ideas about frameworks not only for artificial but also for natural (human's) intelligence. In Chapter 12, Vladimir Rybakov proposes a framework of temporal logics for dealing with chance discovery. In chance discovery, the causality of events before and after a chance is uncertain, and only partially observable — the causal logics behind them cannot be completely possible to explain and are hierarchically structured from local sequences and global scenarios. Rybakov proposes a decidable logic of which the satisfiability is solvable, where operations of the linear temporal logic are combined with introducing operators corresponding to uncertainty temporally local and global discoveries. This will be not only a basic of computation but a starting point to discuss in what logical framework we should discuss chance discovery. In Chapter 13, Ogaard and March focus on pilot behaviors taken from aircraft telemetry data collected by a mobile ground-based sense-and-avoid system for Unmanned Aircraft System (UAS) with sensors and customized visualization software. In order to estimate the current risk of midair collision, they developed probabilistic models for the behavior of pilots of manned aircraft. Complex subpaths were discovered from the data using an ant colony algorithm, and probabilistic models were mined from those subpaths by extending existing algorithms. Signs of failures may be detected objectively by the method, and we may also expect a metacognitive effect i.e., pilots may find mental subpaths that are latent in their own cognitive process which may lead to unexpected risky behaviors. In Chapter 14, Hidenao Abe and Shusaku Tsumoto presents a method for characterizing and clustering a large number of documents, considering the features of temporal changes in the terms appeared in a document collection. They experimentally obtained essential features in the temporal behaviors of four conference series on data mining, from corresponding published document collections.

An extracted change may turn out to be a chance, if applied to documents of one's interesting domain. Finally in Chapter 15, Yoshiaki Okubo, Makoto Haraguchi, and Takeshi Nakajima presents a method for finding an *indicative pattern*, that is a rare but noteworthy pattern — an itemset consisting of several general items but has a small degree of correlation. Then they discuss the relationship of indicative patterns and chance patterns , that is their original abstraction of KeyGraph? which had been created by Ohsawa. A chance pattern is an indicative pattern supported by a pair of more frequent base patterns *BL* and *BR*. Such a pattern is expected to imply some hidden relationship between *BL* and *BR*, as in the way KeyGraph® visualized bridges between frequent patterns as in Fig. 1. They conclude that it might be required to take a causal relationship between a chance pattern and a base pattern into account so that a chance pattern can actually work as a valuable trigger for chance discovery.

In summary, we can point out three major progresses in chance discovery since our last book on chance discovery [Ohsawa and Tsumoto, 2006]: First, based on our common sense that chance discovery is the multidisciplinary research field based on at least two established domains (cognitive science and data mining with visualization), we are involving fresh colleagues from application domains and basic technologies and sciences. Application domains include health care, aircraft control, energy plant, management of technologies and finances, product designs/innovations, marketing, etc, and basic technologies and sciences include sensor technologies, medical sciences, etc. Second, both cognitive scientists and computational scientists who interacted with us deepened and extended their fruits — they are too deep and wide to follow up in this narrow space. Third, time came to be considered and introduced explicitly as a significant variable ruling the causality from background situations to chances and from chances to its impacts on events and actions of humans in the future. Readers may urge us to list the fourth, fifth, sixth,... progresses and we may be able to do so, but let us stop here.

As editors we expect the readers will involve colleagues, living in the real world of human's decision making, to have them join our future activities — although we do not aim to be a large community but rather seek to stay compact enough to continue meaningful synthesis of thoughts via curations of chances and intimate communications.

May 2012 Yukio Ohsawa
Tokyo Akinori Abe

References

[Ohsawa and McBurney, 2003] Ohsawa, Y., McBurney, P. (eds.): Chance Discovery. Springer (2003)

[Ohsawa and Usui, 2005] Ohsawa, Y., Usui, M.: Workshop with Touchable KeyGraph Activating Textile Market. In: Abe, A., Ohsawa, Y. (eds.) Readings in Chance Discovery, Advanced Knowledge International (2005); note: The experiments for this paper were conducted by 2001

[Ohsawa and Tsumoto, 2006] Ohsawa, Y., Tsumoto, S. (eds.): Chance Discoveries in Real World Decision Making, Data-based Interaction of Human Intelligence and Artificial Intelligence. SCI, vol. 30. Springer (2006)

Contents

Cognition and Communication toward Chance Discovery

Curation in Chance Discovery 1
Akinori Abe

Turning Down a Chance: An Argument from Simplicity 19
Emanuele Bardone, Lorenzo Magnani

A Chance Favors a Prepared Mind: Chance Discovery from Cognitive Psychology 33
Hitoshi Terai, Kazuhisa Miwa

Data Visualization as Chance Curation

Chance Discovery with Self-Organizing Maps: Discovering Imbalances in Financial Networks 49
Peter Sarlin

Map Interface for a Text Data Set by Recursive Clustering 63
Wataru Sunayama, Shuhei Hamaoka, Kiyoshi Okuda

Multimodal Discussion Analysis Based on Temporal Sequence 79
Katsumi Nitta

Framework of Early Adopters Roaming among Tribes for Discovering Innovative Creation 91
Mu-Hua Lin, Chao-Fu Hong, Hsiao-Fang Yang, Ai-Ling Wang

Data-Driven Innovation Technologies for Smarter Business from Innovators' Market Game to *iChance* Creativity Support System 107
Hao Wang, Yukio Ohsawa

Computational and Logical Cutting Edges for Analysis and Synthesis of Data

Paired Evaluators Method to Track Concept Drift: An Application in Finance 127
Masabumi Furuhata, Takanobu Mizuta, Jihei So

Efficient Service Discovery among Heterogeneous Agents Using a Novel Agent Ranking Algorithm 143
Hao Lan Zhang, Clement H.C. Leung, Chaoyi Pang, Xijin Tang

Discovering the Chances of Health Problems and Falls in the Elderly Using Data Mining 163
Bogdan Pogorelc, Matjaž Gams

Temporal Logics Modeling Logical Uncertainty, Local and Global Chance Discovery 177
Vladimir Rybakov

Discovering Probabilistic Models of Pilot Behavior from Aircraft Telemetry Data 191
Kirk Ogaard, Ronald Marsh

Constructing Feature Set by Using Temporal Clustering of Term Usages in Document Categorization 215
Hidenao Abe, Shusaku Tsumoto

Finding Rare Patterns with Weak Correlation Constraint: Progress in Indicative and Chance Patterns 231
Yoshiaki Okubo, Makoto Haraguchi, Takeshi Nakajima

Author Index 247

Curation in Chance Discovery

Akinori Abe

Faculty of Letters, Chiba University
Chiba, Japan
ave@ultimaVI.arc.net.my

Abstract. In this paper, a definition and function of a curation are extended according to the chance discovery aspects. A new definition, function, and effect of a curation in chance discovery are discussed. First, ordinal types of curation are reviewed. Definition by American Association of Museums Curators Committee (AAMCC) and Digital data curation are shown. Especially the latter is not curation in (art) museum but for digital data. In addition, interesting (from the viewpoint of chance discovery) curatorial cases are introduced. Based on those definitions and applications, a new type of curation in chance discovery is discussed. Our previous researches are also discussed from the viewpoint of curation. At the end of this paper, new type of curarion is discussed from the aspect of affordance and communication.

Keywords: abduction, chance discovery, curation, affordance, communication.

1 Introduction

A display strategy sometimes plays a significant role. For instance, I found an interesting situation in a market store in Japan, where meats (chicken) and asparagus were sold together. The other vegetables were not sold there. In a general sense, asparagus should be sold in a vegetable selling zone. However, in the market store asparagus were sold in both meat and vegetable selling zone. I did not ask the staff in the market store the reason of their strategy. I think it might have produce potential or hidden needs and desires for customers. Several customers might have bought both meat and asparagus to cook inspired by an implicitly suggested recipe, which would be asparagus rolled with chicken. The reverse activity is sometimes referred to as a sample case in basket analysis [8]. It might be a result from such activities. In addition, in [25], Shimada reports that in a vegetable selling zone, with smell of curry by a special machine, potatoes and carrots are apt to be sold well. The above strategies are regarded as intended combinations of different or unexpected things which will produce (probably) intended effects on customers.

Similarly, curators in (art) museums usually design exhibitions to educate or show their concepts or thoughts to audiences from certain aspects. For that they select or borrow collections, and plan to place the collections. And usually,

Y. Ohsawa & A. Abe (Eds.): Advances in Chance Discovery, SCI 423, pp. 1–18.
springerlink.com

in such exhibitions collections are exhibited in chronological order or grouped in the same or similar categories. For instance if a curator want to build an exhibition whose title is "American Stories: Paintings of Everyday Life 1765–1915[1]," paintings drawing everyday life will be collected or borrowed. Painters to be collected or borrowed are Thomas Eakins, John Singer Sargent, and Mary Cassatt etc. And those collections were placed in chronological order. This is a very understandable example. More complicated exhibitions will also follow the same or similar procedure.

Actually, those strategies are conducted according to curators' own inspirations, and it will be successful if receivers (audiences) are controlled as curators expected. However, sometimes audiences find the other meanings which curators did not expect. From the viewpoint of chance discovery [22], it will be more successful for curators.

In chance discovery a chance is defined as "*a novel or rare event/situation that can be conceived either as an opportunity or a risk in the future*" [22]. That is, it can also be regarded that a chance is an alarm or a symptom like an inflation of money supply or a big difference between future (estimated, reserved) and current stock prices that will change the middle or long term economic "bubble" situation (Japan, in 1990). In addition, of course, a chance can be used for positive situations.

Thus it is very important to offer opportunities where receivers can feel and obtain chances in various situations. Many applications on chance discover have been proposed in these 10 years [4,23]. For instance, visualization systems for making users aware of unconscious preferences [10,17], an analogy game which varies a construction of concepts according to perceptions, categorizations, and areas of focus derived from the expertise of the observer [21], a deposit overflow determination system to prevent various financial crises [28], ISOR-2, a combination of case-based reasoning and statistical modeling system which can deal with medical exceptions [27], and a web-based interactive interface which can check hidden or rare but very important relationships in medical diagnostic data sets [5] have been proposed in [4]. Those applications are real world applications where a discovery of chances plays an important role. However, strategies how to display chances have not been explicitly discussed in many applications.

The above example in a market store can also be regarded as a type of chance discovery application, because it generated a hidden or potential purchase chance to customers. Customers who were inspired by the combination of chicken and asparagus would have bought either or both of them for a dinner. Actually, this is not a task in museums, but it can be also regarded as a curator's work (curation).

In this paper, a definition and function of a curation are extended according to the chance discovery aspects. A new definition, function, and effect of a curation in chance discovery are discussed.

---

[1] A special exhibition at the Los Angeles County Museum of Art (December 19, 2009–July 25, 2010).

2 Curation

This section reviews various types of curation. Actually "curatorial task" is usually performed for tasks in (art) museum. Many cases introduced in this section are curatorial works in gallery and (art) museum. In addition a new type of curation, digital data curation, is also reviewed.

2.1 (General) Curation

There is at least a person who is responsible as "curator" in (special) exhibitions, galleries, archive, or (art) museums. Their main task is a curatorial task, which is multifaceted. Curator comes from a Latin word "cura" which means cure. Then originally it used for a person who take care of a cultural heritage.

In the report by American Association of Museums Curators Committee (AAMCC) [9], they pointed out "curators are highly knowledgeable, experienced, or educated in a discipline relevant to the museum's purpose or mission. Curatorial roles and responsibilities vary widely within the museum community and within the museum itself, and may also be fulfilled by staff members with other titles." Then they showed the definition of curator as follows;

- Remain current in the scholarly developments within their field(s); conduct original research and develop new scholarship that contributes to the advancement of thc body of knowledge within their field(s) and within the museum profession as a whole.
- Make recommendations for acquiring and deaccessioning objects in the museum collection.
- Assume responsibility for the overall care and development of the collection, which may include artifacts, fine art, specimens, historic structures, and intellectual property.
- Advocate for and participate in the formulation of institutional policies and procedures for the care of the collection that are based on accepted professional standards and best practices as defined by AAM, CurCom, and other relevant professional organizations.
- Perform research to identify materials in the collection and to document their history.
- Interpret the objects belonging or loaned to the museum.
- Develop and organize exhibitions.
- Contribute to programs and educational materials.
- Advocate and provide for public use of the collection.
- Develop or contribute to monographs, essays, research papers, and other products of original thought.
- Represent their institution in the media, at public gatherings, and at professional conferences and seminars.
- Remain current on all state, national, and international laws as they pertain to objects in the museum collection.

In addition, AAMCC showed curatorial responsibilities as follows;

A: Research, Scholarship, and Integrity
B: Interpretation
C: Acquisition, Care, and Disposal
D: Collection Access and Use
E: Replication of Objects in the Collection

Thus curators have responsibilities for various aspects of exhibition activities. However, the most important activity will be a plan of exhibition. For that the above activities such as research, interpretation and acquisition are necessary. They should properly exhibit a truth which is result of their researches and interpretations.

2.2 e-Science Data Curation

The above curation is for actual museums. That is, curation is conducted mainly for actual art works. However, curation in this section is for digital data. There are several differences between digital curation and analogue curation.

JISC pointed out an importance of curation as "promoting good curation and an information infrastructure to capitalise upon and preserve expensively gathered data means bringing together varied technical and managerial resources, and managing these over time. This activity needs to be supported by clear strategies for resourcing and support [16]." They compare curation with archiving and preservation.

- Curation: The activity of managing and promoting the use of data from its point of creation, to ensure it is fit for contemporary purpose, and available for discovery and re-use. For dynamic datasets this may mean continuous enrichment or updating to keep it fit for purpose.
- Archiving: A curation activity which ensures that data is properly selected, stored, can be accessed and that its logical and physical integrity is maintained over time, including security and authenticity.
- Preservation: An archiving activity in which specific items of data are maintained over time so that they can still be accessed and understood through successive change and obsolescence of technologies.

That is, they pointed out that curation is more creative task. Then they showed aspects of curation as follows:

- Trust: Trust can be enhanced by the existence of qualified domain specialists who curate the data.
- Utility: Certain information about the data —where it came from, how it was generated, for example— is necessary to enable future users to gauge the utility and reliability of the data, and indeed any annotation of the data. Data utility also depends on the ability of users to manage and analyse it; data mining tools and algorithms, visualisation tools, user interfaces and portals will play a crucial role in accelerating research.

- Discoverability: How will future users find data, in particular data they do not know exists, in other domains, or archived according to terminology which has fallen out of use? Data access is often organised through portals; how will those portals be organised? What tools will users need to read or use the data, and who will provide these tools?
- Access management: A significant proportion of data involves confidentiality issues. Ownership and rights management also need to be taken into account.
- Heterogeneity: Not only is this data revolution creating a deluge of data, the data itself comes in very many different and often specialist formats, some created by the researchers themselves.
- Complexity: The data can be composite in nature, with links to external objects and external dependencies (such as calibration information), and be highly complex in structure. This complexity represents a significant challenge for the preservation of data.

They use "data curation" because they think data have value. Not only for keeping data but also usability of data for the public, they use the word "curation." Actually, most of data are neither art works nor archaeological artifacts. However, is is important to view data from the aspect of what should be preserved. The main difference between data and art works or archaeological artifacts is that data do not have a shape and cannot exist alone. It is necessary to prepare a container such as a cdrom and a hard disc drive system. Therefore for data curation, "Discoverability" plays a significant role.

2.3 Exhibition "Bacon and Caravaggio"

An exhibition "Bacon and Caravaggio" was held in Museo e Galleria Borghese, Roma, Italy during October 2 2009 and January 24 2010.

The display policy of this exhibition is rather different from that of general special exhibitions. First, the special exhibition was not separated from the space for permanent collections. Of course several Caravaggio's works were exhibited in their original places. The other Caravaggio's works and Bacon's works were exhibited between permanent collections. For this type exhibition, usually an exhibition is educational and two painters are compared in various viewpoint, for instance days and society painters lived. Before arriving at the exhibition, my expectation was that it would be an exhibition to address the contrast between the drawing policies of Bacon and Caravaggio. Caravaggio usually painted a perfect body of human beings. On the other hand, Bacon usually painted a flesh of human beings most of whose parts are removed to express the essence of human existence. However, in the catalogue of the exhibition "Bacon and Caravaggio," at first, Coliva wrote "This exhibition proposes a juxtaposition of Bacon and Caravaggio. It intends to offer visitors an opportunity for an aesthetic experience rather than an educational one... [11]." Then Coliva continued

> "An exhibition of generally conceived and prepared with a historicist mentality, but when it materializes, the simultaneous presence of the

> works — in the sense precisely of their hanging — opens up parallels and poses very complex and spontaneous questions, which may even be unexpected and not all stem exactly from questions initially posed by art-historical motives and theses. There are parallels that appear by themselves to the visitor's sensibility and are not imposed by a theory of the curator. This is certainly one aspect of the vitality of exhibitions, which make the works live and in this are necessary for the works. The display itself, in the sense of the presentation of the works that appear in an exhibition —the spectacle of their being on display — creates trains of thought that are independent of the interpretations provided by art-historical scholarship. And since for a profound experience of understanding a work these ramifications sometimes are more surprising and significant than the achievements of a specialized scholarship in its own field of action, an art raised to the status of an enigma like Bacon's seems to require the gamble of provoking these parallels. And since at the time, and again because of its qualitative greatness, Caravaggio's art deserves a similar provocation, the juxtaposition thus satisfies a legitimate aesthetic desire. On the other hand, the juxtaposition is a modest and prudent solution, not so much for demonstrating, but for offering the attribute of "genius" — which the expressive common language attributes to the great artist of the past — opportunities to manifest itself. And the juxtaposition is induced by the Galleria Borghese itself, one of the most sensitive spaces with the simultaneous presence of genius."

Besides the importance in aesthetics and philosophy, I think the most important point is that "There are parallels that appear by themselves to the visitor's sensibility and are not imposed by a theory of the curator." That is, though actually a curator has a certain philosophy, he/she does not insist his/her philosophy but audiences will be able to discover additional meanings as well as the curator's intended philosophy.

2.4 Joseph Cornell / Jiři Kolàř

From April 19 to May 26, 2007, Pavel Zoubok Gallery, NYC, USA organized an exhibition which combined Joseph Cornell and Jiři Kolàř. Where I have also experienced the same situation as shown above.

In fact, both artists are collagists, but before visiting the gallery, I had not expect such a combination. However the combination of well-known artists was very new and impressive and gave me additional perspective to the art.

Pavel Zoubok Gallery presented the combination based on the following concept. "In bringing these artists together we are confronted with two distinct traditions, one rooted in the fantastical visions of American Surrealism during the 1930s and 1940s and the other in the more politically charged spheres of the Central European avant-garde of the 1950s and 1960s, marked by social and cultural repression. Cornell's world, both inside and outside of the box, is one drawn primarily from the imagination. The worldliness and wonder of his art concealed a reality that was often fraught with sadness and an inability to connect directly with the world beyond Utopia Parkway and nearby Manhattan. By contrast, Jiři Kolàr's life and work reflected economic and political struggle and years spent in exile from his native Prague. The myriad collage techniques that he pioneered over fifty years formed an alternative language at a time when the artist/poet saw those in power employing words as an instrument of oppression and misinformation. This progressively led him to a purely visual means of expression."

In addition, Mullarkey pointed out "While Cornell remains an icon of 20th century art, Kolàr's profile as a collagist has diminished. This show suggests reason for the loss of momentum. Viewed side by side, Cornell stands as the more enduring of the two. Kolàř's work is smart, modish and fastidiously crafted. But Cornell's totems of enchantment achieve a disquieting beauty that transcends their moment."

Thus both artists have quite different social background and culture. It might be better to prepare such knowledge before enjoying the exhibition. However for some persons, it was not necessary to prepare such previous knowledge. Sometimes the previous knowledge is harmful because audiences might stop their thinking. The gallery did not explicitly prepare such information in the gallery. Without any previous information, audiences could enjoy the differences and similarities of those two types works from their own viewpoints.

Thus for me it was a simple juxtaposition of two different artistic things. Even such a situation, audiences would have discovered new or unintentional meanings from the juxtaposition. This type of juxtaposition can offer chances. In the above case, a "juxtaposition" itself can be a curation as a chance discovery application.

2.5 Exhibition in the Museum of University of Tokyo

In the museum of University Tokyo, they tried a unusual and tricky exhibition style. Where no panel for explanation is displayed. A director (curator) Endo pointed out that "We have intendedly organized a space without introduction, information, and educational objective. When audiences watch dead bodies which used to have lives and activities, they will conceive an importance of lives [29]."

In the above exhibition, small number of explanatory panels have been provided. However, in this exhibition, no explanatory panels is prepared. Actually, for (archaeological) museums specialized information will be necessary as basic knowledge or educational knowledge. However sometimes reading such panels requires much times to audiences and removes opportunities of deep understandings of exhibitions from audiences. In the worst case, audiences do not watch displayed things but only read explanations. Thus properly few information gives audiences a chance of thinking and deep understandings of the exhibition.

Thus as shown above, originally a curation was an activity for offering an explicit education to audiences. However a contemporary curation offers audiences a certain freedom or opportunity such as deep thinking and new discovery, which can be regarded as a chance. Sometimes a situation without information or with a few information offers us chances which might become important factors for our future. Curation should be performed with considering such implicit and potential possibilities. In addition, such possibilities should be rather easily discovered and arranged according to the user's interests and situations.

3 Curation as Chance Discovery

3.1 Chance Discovery

Though in various articles, the definition of a "chance" is described which was introduced by Ohsawa [22], I wish to introduce it here again. In fact, it rather differs from the original definition in [22] to reflect the recent research interests.

> A chance is rare, hidden, potential or novel event(s) / situation(s) that can be conceived either as a future opportunity or risk.

Then "chance discovery" research is a type of research to establish methods, strategies, theories, and even activities to discover a chance. In addition, it aims at discovering human factors for chance discoveries. Therefore not only researchers in computer science and engineering but also researchers with different expertise such as psychologists, philosophers, economists and sociologists take part in chance discovery research.

Thus it is very important to offer opportunities where receivers can feel and obtain chances in various situations. Many applications on chance discover have been proposed in these 10 years [4,23]. For instance, visualization systems for making users aware of unconscious preferences [10,17], an analogy game which varies a construction of concepts according to perceptions, categorizations, and areas of focus derived from the expertise of the observer [21], a deposit overflow determination system to prevent various financial crises [28], ISOR-2, a combination of case-based reasoning and statistical modeling system which can deal with medical exceptions [27], and a web-based interactive interface which can check hidden or rare but very important relationships in medical diagnostic data sets [5] have been proposed in [4]. Those applications are real world applications where a discovery of chances plays an important role. However, strategies how to display chances have not been discussed in many applications. Strategy for discovering chances is of course important. In addition, strategy for an easy discovery interface of chances is more important. The above interface based application can be classified to curation type applications.

3.2 Is Curation Chance Discovery?

My experience in a market store can be regarded as a type of chance discovery application, because the strategy generated a hidden or potential purchase chance to customers. Customers who were inspired by the combination of chicken and asparagus would have bought either or both of them for dinner. Actually, this is not a task in museums, but it can be also regarded as a curator's work (curation). Because the strategy includes philosophy in a combination of items, and based on the philosophy it will offer certain effects or education to audiences. Visualization strategies such that referred to above function as curation. Because they display candidate chances in a manner where important or necessary items or events can be easily or interactively discovered by the user.

Thus a new definition of curation in chance discovery is:

- Curation is a task to offer users opportunities to discover chances.
- Curation should be conducted with considering implicit and potential possibilities.
- Chances should not be explicitly displayed to users.
- However, such chances should rather easily be discovered and arranged according to the user's interests and situations.
- There should be a certain freedom for user to arrange chances.

For curatorial task, a serious problem is pointed out. Magnani and Bardone introduced an idea of chance-faking as a possible outcome of the activity of chance-seeking [18]. They discussed the problem by illustrating the idea of bullshit introduced by Frankfurt [12] as an activity promoting fake chances. Compared with lie, they illustrate the problem in bullshit. Since a lie is not informatively empty, people have various mechanisms for detecting lies. On the other hand, a bullshit (fake) can be a semantic attack which is concerning with the manipulation of the meaning a person assign to something that he is going to use in his/her decision-making process. That is, a bullshit has no intention to cheat and be logically true. Accordingly, it is rather difficult to determine it as a fake. It will be necessary to be careful of such chance-faking situation or provide a mechanism to detect such chance-faking situation in curation.

4 Possibility of Curation in Chance Discovery

In this section, I discuss possible curatorial strategies. Actually, all applications were previously proposed by me or us.

4.1 Information Offering Strategies for Dementia Persons

In [3], based on abduction, I formalized a concept of affordance based support system for dementia persons. For a proper and an extended usage of a thing, it is necessary to present proper information of it. However, for dementia persons, ordinal information offering strategies cannot function. Therefore, it is necessary to prepare proper information offering strategies especially for dementia persons. However, for a progressive and promising system, it is not realistic to prepare all the necessary information to things or events. Sometimes such information is not always correct and may change in the future. For instance, it is ridiculous to attach a sign such as "You can sit here." to tree stumps. Instead it is rather realistic to suggest information about its hidden functions. Such hidden information can be presented as certain stimuli in such situations. Because, even for dementia person,if he/she receives certain stimuli, he/she sometimes achieve better performance. The problem is that what type of stimulus will be better to present and how to make it recognize.

For that I introduced the concept of affordance which was ecologically introduced for perceptional phenomena by Gibison [13,14]. Gibson defined the affordance of something as “a specific combination of the properties of its substance and its surfaces taken with reference to an animal.”

Actually such stimulus should be “afforded” by the user. That is, it can be regarded as an “affordance” in an environment. Accordingly we introduce concept of affordance to a dementia care system. Proper affordance might give a certain support to dementia persons understanding (meanings of) objects. When meaning is fixed, by using abduction framework, the affordance determination situation will logically be described as follows:

$$Object \cup affordance \models meaning \quad (1)$$

$$Object \cup affordance \not\models \Box \quad (2)$$

That is, affordance can be regarded as a hypothesis set. We can select consistent affordance (equation (2)) in the environment (hypothesis base) to explain meaning. In addition, for understanding subset of or similar afforded objects ($Object'$), the affordance determination situation will be logically described as follows:

That is, affordance can be regarded as a hypothesis. We can select consistent affordance (equation (2)) in the environment (hypothesis base) to explain meaning. In addition, for understanding subset of or similar afforded objects ($Object'$), the affordance determination situation will be logically described as follows:

$$Object \cup Object' \cup M \cup affordance \models meaning \quad (3)$$

M is a mapping function [15] from $Object$ to $Object'$. That is, to understand the same meaning of the subset of or similar afforded objects, an additional mapping function M is required. Thus if M can be determined and the usage of $Object$ is known, $Object'$ can also be understood. In fact, for normal persons, M is easy to understand. However, for dementia persons, it is pointed out that it is rather difficult to understand and determine M.

The most important issue is how to suggest hidden information as affordance. My assumption is that complex situation can be transformed to a combination of simple situations. This type of combination can be achieved as curation. For such system, chance discovery based curation strategies should be introduced to offer understandable mapping

4.2 Decision Making Considering Value or KANSEI

I introduced the concept of “value” to the abduction procedure [7] and proposed a new strategy for the “recommendation” or decision making. Where abduction is extended to deal with potential values or KANSEI[2] which might be discovered by an interactive procedure. I introduced concept of hypothesis expiration, where hypothesis has a certain parameter to show its status —how many time units have passed since it was generated? In addition, I introduced a concept of "weak nogood." This type of nogood is not always considered. That is, sometimes or according to the request, nogood relationship can be cancelled. For instance, usually we do not wear silk ware and Uniqlo[3]'s jeans. Then the relationship can be nogood, but it is not so critical relationship.

$$Cloth\_up(\_, \_, silk), Cloth\_bottom(Uniqlo, navy, jeans)$$
$$\text{:- } nogood(weak). \tag{4}$$

Therefore, the final evaluation can be asked to the user. Then for value sensing, I introduce an additional rule set for hypothesis evaluation. This type of rule deals with value or so called KANSEI (affective or emotional aspect of human activity). It is not necessary to be evaluated during the main abduction procedure. Since it describes the relationships between hypotheses and the (affective) effects from the hypotheses, it does not have any influence to the inference mechanism or results in the main abduction. It deals with rather additional or supplemental information. For instance, it is described as shown below:

$$Cloth\_up(Uniqlo, \_, \_) \text{ :- } casual. \tag{5}$$

$$Cloth\_up(\_, \_, silk) \text{ :- } formal. \tag{6}$$

Abduction adopting the above rule sets explains an observation and generates hypothesis sets as candidate recommendation. For instance, instead of a weak constraint (4), a combination of $Cloth\_up(Versus, gold, silk)$ and $Cloth\_bottom$ $(Uniqlo, navy, jeans)$ can be selected according to the user's KANSEI.

In our framework, the influence from the combination of hypotheses is previously unknown to the user and after a certain interaction, the user will be aware of the additional information for decision making. Accordingly, the user might be able to discover new value during interactions.

I proposed only an abduction framework for recommendation task. Thus I did not propose any explicit strategy for curation, but this type of value generation can be partially regarded as curation. However, more sophisticated strategy will be necessary.

4.3 Interactive Interface for Medical Diagnosis Support

For medical data check and diagnosis, we proposed an interactive interface for medical diagnosis [5]. By using the interface, the user can interactively access a

[2] KANSEI is a Japanese originate concept. KANSEI is defined as "a high-order function of the brain, including inspiration, intuition, pleasure and pain, taste, curiosity, aesthetics, emotion, sensitivity, attachment and creativity [19]".

[3] Japanese company which produces and sells casual wears all over the world.

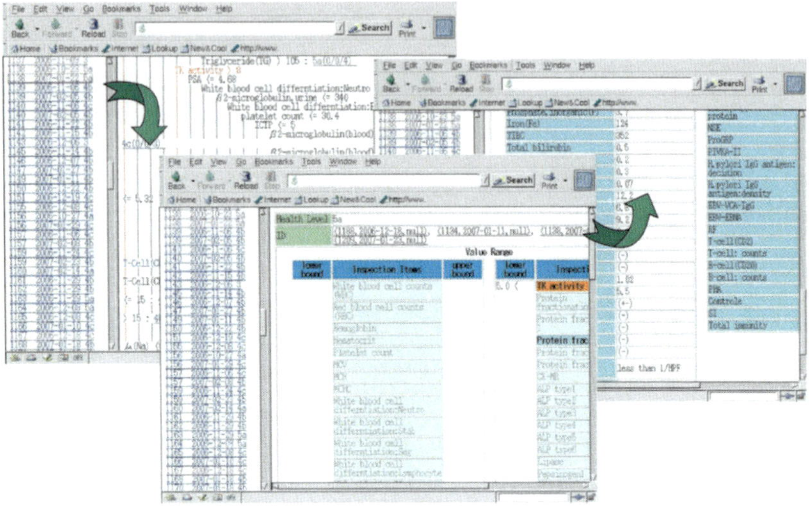

Fig. 1. Interactive interface for medical diagnosis support

certain patient's data. In addition, the system determines patient's health status by applying the learned medical knowledge base.

During the procedure shown in Fig. 1, the user can review a status of a certain patient whose health status is estimated as 5a. In the right window of the right browser, the user can check the selected data to confirm if the patient is in the health status of 5a[4]. Similarly, the user can view all the estimated health status at a glance and check doubtful or uncertain patient's data. Thus the proposed system offers a flexible interface to conduct medical diagnosis and data compilation. Beside, the most important point of the proposed system is that all windows (interfaces) are automatically generated according to the data set and the analyzed results.

By using the interactive interface, the user can discover hidden or potential and important factor of patient. From this viewpoint, we can say this application is chance discovery application. In addition, we can say it can offer curatorial situation.

5 Curation and (Advertising) Communication

In the previous section I pointed out the importance of non-educational curation for assisting chance discovery. I introduced some successful curations as chance discovery for (art) exhibitions in Section 2. However, such successful curations are rather rare. Usually, curations are based on an educational system and it is rather difficult to discover rare issues. Such systems generally provide fixed information and therefore by following such information, we can reach the correct and intended goal. Usually it is happy for us. On the other hand, if we do

[4] In [5], I described health status as health level. For the definition of health level, please refer to [5] etc.

not follow the information, we cannot reach the correct goal, and we will feel unhappy.

5.1 Advertising Communication

Advertising communication also aims at offering certain information to audiences. Sometimes such information is implicitly expressed in an advertisement. As Pop summarized [24], advertising communication relies considerably on inferences and assumptions which help in proceeding towards eventual interpretations. Based on Grice's seminal theory of cooperative communication (cooperative principles, CP) and inferencing through a maxim of "filling in" or/and flouting, different interpretations could be accommodated by the linguistic theory. Pop extended her discussion by introducing Relevance Theory (RT)[26], which explains hidden and additional information in advertisement.

Both curation and advertisement intend to deliver a certain concept to audiences. As easily guessed "communication" plays a significant role in curation and advertisement.

5.2 Advertising Communication and Curation

I will not discuss more details of advertising communication in this paper due to its length limitations. Several discussions on advertisement techniques which are based on language sense techniques were provided in [1]. These were not discussed from the perspective of affordance or abduction, but of advertising communication. Frequently an advertisement planner tries to hide an actual intention in phrases or pictures used in advertisements. For instance, the advertisement shown in Fig. 2 includes a phrase "The Best Things in Life are Basic." This phrase will be based on the proverb "The best things in life are free." By using the strategy, audiences can understand additional meaning from the advertisement.

This type of strategy can be discussed from the aspect of communication. In addition, it can be said that advertisements offer a type of affordance to be selected by audiences. That is, hidden meaning will be selected as a proper affordance. Of course, such affordance will be effectively selected by a suitable curatorial tasks. Thus a mapping function M shown in the equation (3) will also function between the phrase and the proverb.

In the following I will review chance discovery from the aspect of communication. Actually, the following application is affordance, which can be regarded as communication between human and environment. We communicate with an environment and ecologically selected affordance which will be better or best for our lives.

Fig. 2. Example of quotation from a proverb

6 Communication, Abduction and Affordance

In the previous section, I briefly introduced my previous work based on affordance which can also be discussed as an application of curation. For the application, the most important issue is how to suggest hidden information as affordance. My assumption was that complex situation can be transformed to a combination of simple situations. This type of combination can be achieved as curation. For such system, curatorial strategy should be introduced to offer understandable mapping

I formalized affordance selection with abduction and analogical mapping. In addition, analogical mapping can also be executed in an abduction framework. An affordance selection is achieved by communication between environment and humans. If it is difficult, it means that communication between environment and humans is missing. For an appropriate affordance selection, it is necessary to perform suitable communication between an environment and an individual. In this section, I will view missing communication as chance and propose a simple strategy to determine a chance.

6.1 Missing Communication as Chance

I have previously analyzed nurses' activity pattern by using $Kamishibai-KeyGraph^{®}$ [2].

When we analyze nurses' activities by applying $Kamishibai-KeyGraph^{®}$, a result shown in Fig. 3 is obtained. For an export of $KeyGraph^{®}$, it may not be so difficult to determine the point of chance. However, usually this type of task is conducted by both a $KeyGraph^{®}$ expert and an expert in the field (nurse, doctor or risk manager in this case). Of course, a good collaboration sometimes produces novel viewpoint which can also be regarded as a chance. However, it is

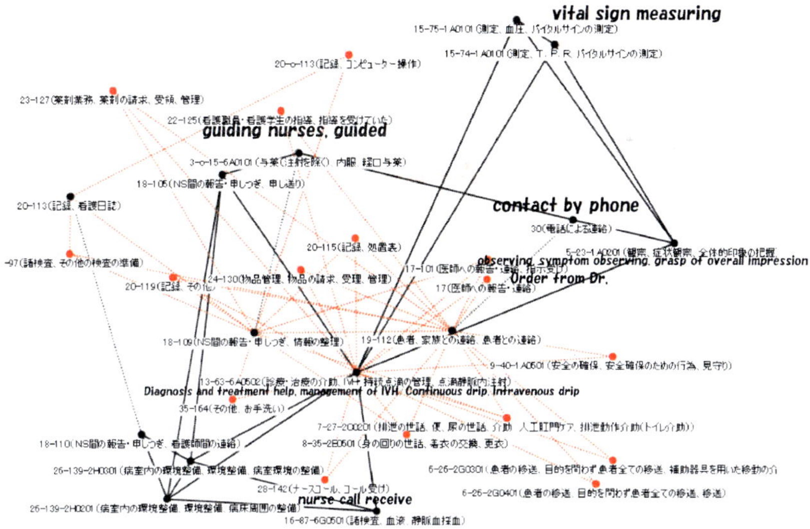

Fig. 3. Result from *Kamishibai* – *KeyGraph*®

not always the case that both experts can work together. The preferable situation is given by a field expert who can properly control such an analysis system to exercise chance discovery.

To go back to the application of affordance, the figure from *KeyGraph*® should include affordances, which the user will be aware of and will select. That is, in the case of Fig. 3, affordances come from several key points which are candidates of chance points as well as the other points. For a sufficient result, users will select proper affordances. For instance, for professional observer, Fig. 3 might be observed as Fig. 4 where light yellow regions contain candidate affordances. For an untrained individual, such an observation will be rather difficult. Therefore suggested markings as those in Fig. 4 will be helpful. However, of course they should be correct and as pointed out before they should not indicate system maker's intentions, which may bias the observer. They should rather be displayed naturally and should not disturb or control the user's thinking process.

6.2 How to Determine Missing Communication?

In [6], we proposed an interactive support system to suggest chance points, which can analyze nurses' conversation to determine points at which communication error might occur. Therefore it is tuned to communication analysis. However, a fundamental concept of the system is based on abduction. That is, the system back-traces communication paths from the point where an accident or incident is noticed.

In general, abduction functions to determine reasons (explanations) for an observation or sometimes for missing events to explain an ideal observation. In

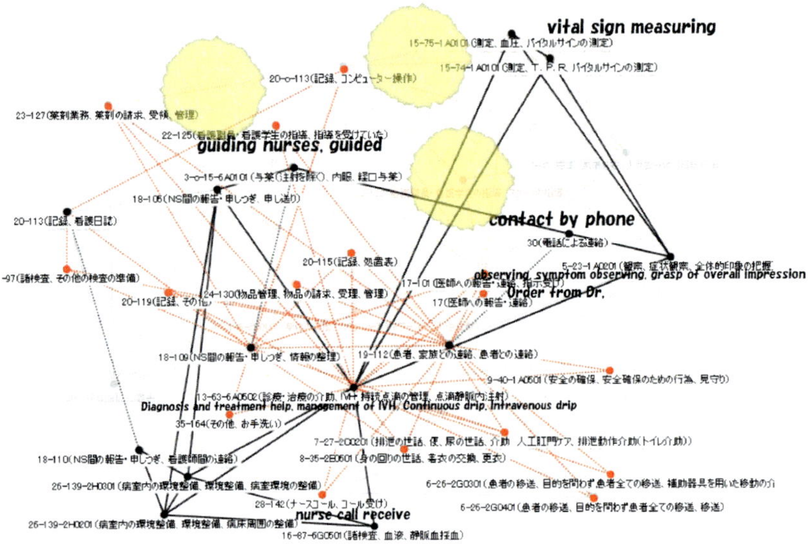

Fig. 4. Result from *Kamishibai – KeyGraph*® with suggestions

fact, as shown above, in the information offering system for dementia patients [3], I applied abduction to generate a mapping for understanding and selecting affordance or obtaining meaning. Thus by adopting the abduction framework, it will be able to suggest missing communication, which can be regraded as a set of affordances. If we adopt a graphical system, for instance *KeyGraph*®, it is necessary to build a set of logical statements from the result. After that abduction can be applied to abduce missing communication, which will be an affordance. Of course, many candidate affordances will be abduced. However, logically certificated candidate affordances will be helpful for the further decision making.

As Magnani and Bardone discussed, the problem of an activity promoting fake chances[5] will usually occur during the activity of chance-seeking [18]. A bullshit has no intention to cheat and to be logically true. Accordingly, it is rather difficult to determine it as a fake, because the above procedure seems to be guaranteed by a logical procedure. In order to ignore such "bullshit," curation will play an important role. That is, by a appropriate curation such "bullshit" will be shown as a lie. For that, curation as communication is very important. During such communication, users will be guided into a proper direction.

In this section, I used an example output by *Kamishibai – KeyGraph*®. Abduction can be applied to various applications, so the above strategy can be applied to any application which has a causation. Thus, a proper strategy based on curation (abduction, affordance, and communication) should be developed for an improved chance discovery.

[5] They call it bullshit.

7 Conclusions

In this paper, I discussed the meaning of curation in chance discovery. I defined "curation" as a task to offer users opportunities to discover chances. In addition, I proposed several features of curation as chance discovery. In fact, the word "curation" is inspired by artistic curation where I have experienced several unexpected impressions in exhibitions. During enjoying such exhibitions, I have began to consider the effect of "curation" in chance discovery. As pointed out in this paper, strategies how to display chances have not been explicitly discussed in many applications. I think it will be time to consider such a curatorial strategy in chance discovery.

In this paper, I have not shown an exact answer to the curation strategies. Instead, I gave several suggestions. For instance, I showed several possibilities for curation in chance discovery by referring to our previous studies. In addition, I view affordance selection from the viewpoint of communication error and curation. Currently, my view for curation seems to tend to a strategy of display or interface. However, for curation, the other strategies can be proposed. In addition, problems such that pointed out by Magnani and Bardone [18] will be solved by means of curation.

References

1. Abe, A.: Language Sense. In: Proc. of PRICAI 2004 Workshop on Language Sense on Computer, pp. 1–9 (2004)
2. Abe, A., Ohsawa, Y., Ozaku, H.I., Sagara, K., Kuwahara, N., Kogure, K.: Communication error determination model for multi-layered or chained situations. In: Proc. of PAKDD 2008 Working Notes of Workshops on Data Mining for Decision Making and Risk Management, pp. 305–316 (2008)
3. Abe, A.: Cognitive Chance Discovery. In: Stephanidis, C. (ed.) UAHCI 2009, Part I. LNCS, vol. 5614, pp. 315–323. Springer, Heidelberg (2009)
4. Abe, A. (eds): Special issue on Chance Discovery. International Journal of Advanced Intelligence Paradigms, vol. 2(2/3) (2010)
5. Abe, A., Hagita, N., Furutani, M., Furutani, Y., Matsuoka, R.: An Interface for Medical Diagnosis Support —from the viewpoint of Chance Discovery. In: [4], 283–302 (2010)
6. Abe, A., Ohsawa, Y., Ozaku, I.H., Sagara, K., Kuwahara, N., Kogure, K.: Communication Error Determination System for Multi-layered or Chained Situations. Fundamenta Informaticae 98, 123–142 (2010)
7. Abe, A.: Abduction dealing with potential values. In: Proc. SMC 2010, pp. 1279–1285 (2010)
8. Agrawal, R., Imielinski, T., Swami, A.: Mining Association Rules between Sets of Items in Large Databases. In: Proc. of the ACM SIGMOD Int'l Conf. on Management of Data (ACM SIGMOD 1993), pp. 207–216 (1993)
9. American Association of Museums Curators Committee: A code of ethics for curators (2009), http://www.curcom.org/_pdf/code_ethics2009.pdf
10. Amitani, S., Edmonds, E.: A Method for Visualising Possible Contexts. In: [4], 110–124 (2010)

11. Colvia, A.: Caravaggio Beckons to Bacon: The Beauty of Sorrow. In: Coliva, A., Peppiatt, M. (eds.) Caravaggio Bacon, pp. 17–22 (2009)
12. Frankfurt, H.: On Bullshit. Princeton University Press (2005)
13. Gibson, J.J.: The Theory of Affordances. In: Shaw, R., Bransford, J. (eds.) Perceiving, Acting, and Knowing (1977)
14. Gibson, J.J.: The Ecological Approach to Visual Perception. Houghton Mifflin (1979)
15. Goebel, R.: A Sketch of Analogy as Reasoning with Equality Hypotheses. In: Jantke, K.P. (ed.) AII 1989. LNCS, vol. 397, pp. 243–253. Springer, Heidelberg (1989)
16. JISC: e-Sciece Data Curation Report (2004), `http://www.jisc.ac.uk/e-sciencecurationreport.pdf`
17. Maeno, Y., Ohsawa, Y.: Reflective visualization and verbalization of unconscious preference. In: [4], pp. 125–139 (2010)
18. Magnani, L., Bardone, E.: Faking Chance: Cognitive Niche Impoverishment. In: Setchi, R., Jordanov, I., Howlett, R.J., Jain, L.C. (eds.) KES 2010, Part III. LNCS(LNAI), vol. 6278, pp. 331–338. Springer, Heidelberg (2010)
19. Ministry of Economy. Trade and Industry (eds.), "KANSEI" Initiative — Suggestion of the fourth value axis– Ministry of Economy, Trade and Industry Press Release (May 2007), `http://www.meti.go.jp/english/information/downloadfiles/PressRelease/080620KANSEI.pdf`
20. Mullarkey, M.: Bound by Time (2007), `http://www.maureenmullarkey.com/essays/kolar.html`
21. Nakamura, J., Ohsawa, Y., Nishio, H.: An analogy game: toward cognitive upheaval through reflection-in-action. In: [4], pp. 220–234 (2010)
22. Ohsawa, Y., McBurney, P. (eds.): Chance Discovery. Springer (2003)
23. Ohsawa, Y., Tsumoto, S. (eds.): Chance Discoveries in Real World Decision Making, Data-based Interaction of Human Intelligence and Artificial Intelligence. SCI. Springer (2006)
24. Pop, A.: Covert Communication in Advertising: A Case Study, `http://www.upm.ro/facultati_departamente/stiinte_litere/conferinte/situl_integrare_europeana/Lucrari2/AnisoaraPop.pdf`
25. Shimada, C.: Hot & Cool, Weekly Toyo Keizai, vol. 6267, p. 32 (June 19, 2010) (in Japanese)
26. Sperber, D., Wilson, D.: Relevance, 2nd edn. Blackwell (1995)
27. Vorobieva, O., Schmidt, R.: Case-Based Reasoning to Explain Medical Model Exceptions. In: [4], pp. 271–282 (2010)
28. Yada, K., Washio, T., Ukai, Y.: Modeling Deposit Outflow in Financial Crises: Application to Branch Management and Customer Relationship Management. In: [4], pp. 254–270 (2010)
29. Miyadai, E.: Possibility of exhibition style. Eevening Asahi Shimbun (March 11, 2010) (in Japanese)

Turning Down a Chance
An Argument from Simplicity

Emanuele Bardone[1,2] and Lorenzo Magnani[2,3]

[1] Institute of Informatics, Tallinn University, Tallinn, Estonia
[2] Department of Arts and Humanities, Philosophy Section and Computational Philosophy Laboratory, University of Pavia, Pavia, Italy
[3] Department of Philosophy, Sun Yat-sen University, Guangzhou, P.R. China
bardone@tlu.ee, lmagnani@unipv.it

Abstract. The notion of bullshit has quite recently acquired a theoretical and philosophical dignity. Described as the careless attitude that an agent has towards the truth-value of what he believes in, bullshitting is a valuable candidate for shedding light on those situations in which a chance should be turned down inasmuch as it is a fake chance. By contrasting the idea of bullshitting with that of scientific rationality, we will contend that the former favors what we call *cognitive baroquism*. That is, bullshitting has the major effect of unnecessarily complicating things and so intoxicating our cognitive niches with fake chances. In order to capture this feature of bullshitting as chance-faking, we will develop an argument from simplicity. Coming from the tradition of philosophy of science, the notion of simplicity might be of great help in spelling out the main effects of chance-faking and, at the same time, it gives us a valuable heuristic to understand when a chance should be rather turned down than actively pursued and implemented.

Keywords: chance-faking, cognitive niche, scientific rationality, cognitive baroquism, bullshit.

1 Maintaining Cognitive Niches and the Problem of Detecting Fake Chances

As defined by Ohsawa and McBurney [19], a chance is a new event or situation conveying both an opportunity and a risk in the future. Recently, a number of contributions have acknowledged the abductive dimension of seeking chances with relation to science [12,15,13,1]. As maintained by Magnani and Bardone [15] and Abe [1], the process of chance detection (and creation) is resulting from an inferential process – mainly abductive – in which the agent exploits latent clues and signs signaling or informing the presence of an action opportunity [15,2]. In this case, as argued by Magnani [13] the abductive inferential dimension has to be considered beyond its sentential/computational one.

Accordingly, an inference is a form of sign activity in which the word sign encompasses several types of sign, for instance, symbol, feeling, image, conception, and other representation [20, 5.283]. Moreover, the process of inferring – and so the activity of chance seeking and extracting – is carried out in a distributed

Y. Ohsawa & A. Abe (Eds.): Advances in Chance Discovery, SCI 423, pp. 19–31.
springerlink.com

and hybrid way [13]. This approach considers cognitive systems in terms of their environmental situatedness: instead of being used to build a comprehensive inner model of its surroundings, the agent's perceptual capacities are seen as simply used to obtain "what-ever" specific pieces of information are necessary for its behavior in the world. The agent constantly "adjusts" its vantage point, updating and refining its procedures, in order to uncover a piece of information. This resorts to the need of specifying how to efficiently examine and explore and to the need of "interpreting" an object of a certain type. It is a process of attentive and controlled perceptual exploration through which the agent is able to collect the necessary information: a purposefully moving through what is being examined, actively picking up information rather than passively transducing [24]. In this sense, humans like other creatures are ecological engineers, because they do not simply live their environment, but they actively shape and change it looking for suitable chances, epistemic for example, like in the case of scientific abductive thinking.

Generally speaking, the activity of chance-seeking as a plastic behavior is administered at the eco-cognitive level through the construction and maintenance of the so-called cognitive niches.

A cognitive niche can be defined as that part of the environment that has been modified so as to better identify, store and preserve profitable chances for solving problems [15]. The notion of cognitive niche encompasses a wide range of human-tailored environments, whose materiality and breadth is not always immediately intelligible. It is a cognitive niche, for instance, the science laboratory, in which the environment comprising various artifacts and instruments is specifically designed for carrying out certain epistemic tasks, say, related to the creation of scientific knowledge.[1] But at the same time that particular cognitive niche is part of a larger one, which is the scientific community as a whole. In such community technological artifacts, tools, but also cultural practices and routines, highly qualified people, and various institutions (universities, centers of research, and governmental organizations) contribute to forming a sophisticated network of interaction supposed to increase our chances to attain certified knowledge.

At a more abstract level, the various cognitive niches humans live in are responsible for delivering those clues and signs informing about an (environmental) chance. So, the mediating activity of inferring as sign activity takes place (and is enhanced) for the presence of the so-called eco-cognitive inheritance system [18]. That is, humans can benefit from the various eco-cognitive innovations as forms of environmental modifications brought about and preserved by the previous generations. Indeed, many chance-seeking capacities are not wired by evolution, but enter one's behavioral repertoire because they are secured not at the genetic level, but at the eco-cognitive one – in the cognitive niches. The second important point to mention is that humans as chance extractors act like eco-cognitive engineers [15,2]. Accordingly, they take part in the process of extracting chances

[1] A broader epistemological analysis of the notion of cognitive niche is provided in Magnani [13, ch. 6] and Bardone [2, ch. 3].

by performing smart manipulation in order to turn an external constraint into a part of their extended cognitive system.

The persistence of cognitive niches is granted by various activities of cognitive niche maintenance whose main role is to preserve the integrity of our cognitive repertoires. So, humans as eco-cognitive engineers are also chance-maintainers [2]. Humans act so as to preserve those chances (and "chances of chances") that have been proved as successful and worth pursuing. That is what allows us to have an eco-cognitive inheritance that, in turn, enriches or even creates the behavioral chances a community or a group of people has for surviving and prospering.

Innovation management can be considered an activity of cognitive niche maintenance, as it deals with how to enrich a pre-existing cognitive repertoire. Technological innovation is indeed part of the process in which our cognitive niches are progressively enriched with new chances at several levels, namely, motor, sensory, and cognitive [2]. However counterintuitive it might sound, not all improvements due to technological innovation are really enhancing. Not all of them can really enrich our cognitive repertoire with new and valuable chances. Some technologies may not be enriching at all. Conversely, they may contribute to introducing chances that intoxicate[2] or poison pre-existing cognitive niches. Such chances are *fake* chances that should be turned down. Now the question is: when should we turn down a (fake) chance, notwithstanding the fact that it is a product of technological innovation?

In investigating chance-faking, we posit that pointing at the very roots of scientific rationality is extremely fruitful. As a matter of fact, the unprecedented success that human beings have achieved in the last centuries as eco-cognitive engineers bears an essential connection to the progress of science and scientific rationality. The latter does not only regard with the results of science; it embraces all those habits – epistemic, but also moral, social, and political ones – which are prerequisites for the development and progress of science in society. Generally speaking, scientific rationality is characterized by all those methods and forms of reasoning that consider truth as the value one should not compromise on. Indeed, it is very important to note that the kind of truth we are talking about has a specific meaning to science and within scientific practice.

We will start our discussion with a philosophical account of bullshitting provided by Harry Frankfurt. If the hallmark of scientific rationality is a strong commitment towards the search for a reliable truth, bullshitting is conversely characterized by the absence of such commitment. The absence of commitment towards truth makes bullshitting the perfect spot from which we can look at the phenomenon chance-faking.

---

[2] Recent philosophical literature on information technology and its role in the birth of the so-called "ignorance society" stresses the attention to the general function of "infoxication" [6]: "a kind of intoxication brought on by too much information which translates into increasing difficulty discriminating between what is important and what is superfluous, and choosing reliable sources of information" [5, p. 25].

By contrasting the notion of bullshitting with that of scientific rationality, we will contend that bullshitting favors what we call *cognitive baroquism*. That means that bullshitting has the major effect of unnecessarily complicating things. In order to capture this feature of bullshitting, we will develop an argument from simplicity. More precisely, we will show that the very notion of simplicity – coming from the tradition of philosophy of science – might be of help in spelling out the main effects of chance-faking.

In the last part of the paper we will apply the idea of cognitive baroquism to technology. We will describe two cases in which bullshitting may affect the way we use technology and, as a result, facilitate the proliferation of fake chances.

2 What Bullshitting Is and Its Philosophical Relevance

The notion of bullshit introduced by Frankfurt [7] may help describe a fundamental feature of people: when they exhibit a carelessness about truth and can easily perpetrate violence – favoring deception and fraud – just thanks to this systematic undervaluing of truth [16]. Again, it is extremely important to note that in this case we consider truth as bounded up with the concepts of reasons, evidence, experience, inquiry, reliability, and standard or rational belief, for example in agreement with a kind of scientific mentality.

According to Frankfurt, there is an important distinction to make between a bullshitter and a liar. The difference between the two is that the liar has a general concern about truth. And this is just because, in order to tell a lie, he has to know what the truth is. Although the liar fails to be cooperative with respect to a certain state of things in the world, he is indeed cooperative with respect to his attitude towards truth. The fabrication of a lie may require a great deal of knowledge and it is mindful: putting it simple, it requires the guidance of truth. More generally, a certain state of mind – namely, an intention to deceive – is required to the liar while making a statement. This attitude is what makes his statement potentially informative. For instance, consider the case of a person telling us that he has money in his pocket, when he has not. His lie is informative as one can guess whether he lied or not. What is interesting about lying is that there always is a reason why a person may not tell the truth: lies and deceit can be detected. People, for instance, have at disposal both verbal and non verbal cues enabling them to detect potentially deceiving situations [25]. A minor detail about the way a person dresses may be a chance suggesting a man that his wife is cheating on him, and vice-versa. Sometimes people fear the consequences of knowing the truth, therefore they prefer not to investigate. But this does not mean that they would not succeed. Quite the opposite. So, given the fact that a liar is committed to the truth-value of one's belief, lying may contribute to hiding a chance, but at the same time it might be revealing it, as one detects the deceiving intention of the liar.

According to Frankfurt, the case of bullshit is different, as the bullshitter is supposed to lack any concern or commitment to the truth-value of what he says. What turns out to be extremely puzzling is not the content, but his attitude. For instance, as we have already pointed out above, a liar voluntarily makes a wrong claim. But in doing so he conveys a certain commitment to the truth-value of what he claims. A bullshitter does not care about it. As just mentioned, a liar has a deceptive intention that can be detected. Whereas the case of bullshitter is different. When a person believes P, he intends to believe P. And that intention becomes meaningful to other people. In the case of bullshitter, he believes without a real intention to believe what he believes. So, what really defines a bullshitter is his attitude towards truth: he fails to be committed to truth. He simply does not care whether what he says is true or false, accurate or inaccurate.

3 Bullshitting as Chance-Faking

The philosophical illustration of bullshit we have presented so far allows us to argue that bullshitters are basically *chance-fakers*. Why are they so? The deceiving character of chance-faking is related to the fact that one acts as if a chance were present, when it is not. Roughly speaking, what comes out of the bullshitter's mouth is *hot air* or *vapor*, meaning that the informative content transmitted is *nil*.

The deceiving character of chance-faking is particularly evident in the case of those cognitive processes involving a collaborative dimension, for instance, the case of second-hand knowledge, that is, the possibility of passing a chance on to another person. But how does the chance-faker engage other people in discussion, for instance?

As argued by Simon, we do lean on what other people say. That disposition to rely on social channels in problem-solving activities is what he called *docility* [21]. People tremendously benefit from aids and resources provided by their fellows. This has a major cognitive advantage: that almost anybody can trust other people and so have at disposal chances that, first of all, he has never personally experienced, and, secondly, that are already available to pick up. That is one of the most important assets describing cognitive economy, that is, the need to reach a sort of trade-off between accuracy of a decision and the limited time one is bounded to. Indeed, trust is not informatively empty. One decides to trust another person, because he has reasons to do so. There is a number of clues we make use of in order to consider a particular source of information (a person, for instance) as trustworthy or not. For instance, people usually tend to trust people that exhibit some authority [11]. In our opinion, the chance-faker basically provides his partner in conversation with chances that are basically faked. To make this point clearer, we introduce the notion of *frame*.

Usually, when entering a discussion a person is supposed to have some expectations. For example, he wants to gain a favor or he wants to display a certain identity. On some other time, he wants to express his dominance over the other or he simply wants to entertain himself or others taking part in the conversation. Usually, all of these reasons lead a person to *frame* what he is doing in a particular way rather than another depending on what his purpose is [8,9]. Now, the question is: what kind of *frame* – to use Goffman's terminology – does the chance-faker select? What does he hold in his mind? What purpose?

Our take is that the chance-faker speaks without planning and doing so he simply *blurts*. As Hample put it, blurting "comes simply out of cognition without alteration" [9, p. 23]. Basically, he acts as if he responded *unreflectively* and *mindlessly* to an invitation to have his say – no matter whether it is real or imaginary. Instead of admitting that he has nothing to say (and hence holding his peace), the chance-faker enters a debate or discussion without questioning if he has actually something to say. It seems that he suffers from a sort of pathological certainty, meaning that he does not come to doubt when he should about his competences and knowledge. So, his pathological certainty may produce what could be called *pathological opining*: he cannot abstain from opining about topics clearly beyond his knowledge.

We have a last remark on frames and bullshitting. As already mentioned, a person entering a discussion usually bears in mind a certain understanding of what is going on. That is, he (more or less tacitly) interprets why others are arguing, i.e. their frame. Indeed, it is important to understand what frame a partner is using during a conversation: it avoids misunderstandings and, upon occasion, facilitates cooperation and information exchange. By contrast, when two or more people do not share the same frame, cooperation is less likely to come about. One can even project a false frame for his partner(s) and then hack the conversation to have a private gain. When this happens, Hample *at al.* argue that the victim "is said to be *contained*" in the swindler's frame [10]. We contend that something similar happens with the chance-faker. We argued that the chance-faker has no interest or purpose in his mind when entering a conversation, but he simply responds to a (sometimes imaginary) invitation like reacting to a stimulus (for instance, "rate this product!"). Hence, he actually does not project any frame but, insofar as he is engaged in a conversation, he gives the impression he is employing a frame. It is worth noting here that the deceptive nature of bullshitting does not rest on the projection of a false frame, but on the fact that his victims are naturally inclined to assume that he actually projects a frame when he does not. The chance-faker's conversational contribution is nil, but nevertheless he is very often taken seriously, say, as if he really contributed.

The major consequence is that an imaginary frame is projected and the chance-faker's victims get into it and thus are *contained*, just like in the case of a fraudulent attack. Indeed, once they realize that the frame was merely imaginary, they may experience all that as fraudulent and aggressive. As a consequence, the chance-faker may even try to justify his misbehavior, for instance, saying that he did not really mean to trick his conversation partners, but that he just wanted to play. This sort of post-projection does not solve the problem. A fake chance is fabricated and it is available to be picked up. In fact the newly-projected meaning sharpens the dissonance felt by his victims. According to Goffman, when a certain activity is re-framed, or *re-keyed*, it is said that a *lamination* (or layer) is added to that activity [8, p. 82]. More precisely, a lamination is a new meaning a person adds up to pre-existing layers. Indeed, the prerequisite for a lamination to be perceived is that the partners in conversation apply the appropriate frame. So, a lamination does not conflict or erase all the pre-existing meanings that a certain activity – for instance, arguing – may have. But it somehow becomes a new layer that metaphorically *laminates* it providing new interpretations, but, at the same time, it also favors the emergence of dissonance, in case the participants are not able to successfully re-key the activity.

4 Chance-Faking as Cognitive Baroquism and Its Application to Technology

As already discussed, bullshitting as chance-faking is a kind of attitude characterized by the absence of commitment towards the truth-value of what one believes in. More generally, chance-faking deals with all those situations in which a person (or a group of persons) displays a careless attitude concerning how to generate and/or select "plausible" and "decent" hypotheses, which eventually gives shape to his beliefs. In this sense, insofar as the chance-faker is involved in generating or selecting hypotheses, he embraces a degraded form of abductive cognition.

The careless attitude toward the whole process of hypothesis generation and selection – which is responsible for belief formation – may be manifest in various ways. Indeed, the most distinctive trait characterizing the poor abductive performance of a chance-faker is that he does not pay attention to evidence.

Consider diagnostic reasoning as a specific case of abductive cognition. A doctor might accidentally produce a fake chance for a patient (for example, the prospect of a surgery), as he settles into a certain diagnosis and subsequently discards all other information, which might disprove it or present other better alternatives. This can be called *pathological opining* [2]. The bullshitter fabricates fake chances by stopping the search, when he *should* continue the search. Which means that he is not able to adopt any form of skepticism. We will come back to this issue later.

There is another aspect – to a large extent ignored – which drastically contrasts, for instance, with scientific rationality: bullshitting – especially in its more sophisticated versions – tends to what we call *cognitive baroquism*.

In belief formation a certain claim may be made more compelling and bolder by other claims that are *auxiliary*. That is, they provide support. Such claims are assumed and they can contribute to fostering our position so that, if they were false, the claim they are supporting would be less compelling or even false. This has a simple consequence: the more assumptions we make, the more knowledge and/or facts we need to corroborate them. Which means that we are far more exposed to getting things wrong. That being said, we claim that chance-fakers do not care about the truth-value of the assumptions they make. And, in doing so, they do not adopt any commitment to simplicity, which, conversely, is peculiar to scientific rationality. We are now going to show how this attitude – the lack of commitment to simplicity – might have negative effect in chance-seeking activities.

Bullshitters fail to exhibit any commitment to the truth-value of what they say, but they successfully manage to say something. Bullshitters always have an answer, no matter what they are talking about, as they rely on what we called pathological opining. By this term we mean that they does not come to doubt when they should about their competences and knowledge: they cannot abstain from opining about topics clearly beyond their knowledge. As mentioned before, they do not adopt skepticism, when they should. Their careless attitude towards any procedure of belief formation and revision allows them to say something about everything *anyway*. In this sense, we posit that chance-fakers *mimic* what philosophers of science call *consilience*.

Consilience refers to how much a theory explains – its explanatory breadth [23]. Generally speaking, the more consilient a theory is, the more it explains. Indeed, the kind of consilience we are talking about is not the same of a scientific theory or hypothesis. Let us say that we rather refer to a *pragmatic* form of consilience than an epistemic one. That is, the ability in a discussion of formulating opinions which allows to jump from one topic to another preserving a certain degree of unity and generality. In this sense, the kind of consilience exhibited by the chance-faker is similar to the one of a *know-it-all*. However, we claim that chance-faker's pragmatic consilience is achieved at the expense of simplicity. We will now illustrate why.

The medieval philosopher William of Ockham pointed out that it is worthless to do with more what can be done with less. In philosophy of science simplicity plays a pivotal role in comparing two or more theories. As defined by Thagard, simplicity refers to the number of kinds of entities (hypotheses, concepts, etc.) that a given theory actually assumes in order to explain certain facts [22]. The less entities a theory assumes, the simpler it is. More generally, simplicity points out that a theory should make use of as less entities as possible. Now, we analogically argue that a chance-faker is completely careless about what he should assume in order his belief to be consistent and epistemically grounded. That is, he can easily resort to pathological opining insofar as he completely dismisses any commitment towards the truth-value of what his beliefs would imply to be acceptable – the auxiliary hypotheses in the case of a scientific theory, for instance. The consequence is the dismissal of simplicity.

What we are going to do now is to try to show how this particular cognitive distortion due to chance-faking may be fruitfully applied to technology. More precisely, we will show that the chance-faking attitude we have just described can also be applied to technology, and so it presents interesting hints about turning down a chance.

As already pointed out, bullshitting is intrinsically related to the way humans form and select their beliefs. But in a way – and this is our major claim – the same attitude might be exhibited with relation to the construction, selection, and evaluation of technologies and the chances they provide. In order to make this clear, we maintain that simplicity and consilience may fruitfully be part of our toolbox for detecting and avoiding what we called fake chances. Indeed, the role of consilience and simplicity for comparing two or more competing scientific theories can only be analogically applied to the case of technology. Clearly, a technology is not a theory, and vice-versa. However, when appropriately modified, the two concepts can be of some interest.

Running the analogy between a theory and a technology, we may say that the (pragmatic) consilience of a technology may refer to the things it enables us to do. Whereas simplicity – we argue – may refer to *the way* we are enabled to do those things. This quite broad statement may be translated into something more specific to technology: that is, simplicity can be related to the auxiliary accessories, technologies, dependencies, but also skills and knowledge a cognitive agent should assume to have in order to exploit all the functionalities of a given technology.

Table 1. Consilience and simplicity in science and technology

	THEORY	**TECHNOLOGY**
CONSILIENCE	Explanatory breadth	N. of functionalities
SIMPLICITY	N. of auxiliary hypotheses	N. of auxiliary accessories

Ideally, a good technology is the one that has many functionalities and, at the same time, that does not require too many accessories, dependencies, etc. To make a trivial example, consider two laptops. They are equipped with the same screen and keyboard size, same weight, same power, etc., but one has a battery lasting 6 hours, whereas the other runs out of battery after 3 hours. The two laptops have the same consilience, but the first is simpler, because we can go around without the need to carry the power adapter and search for an outlet to plug it in. It is more parsimonious in relation with accessories. To make another example, usability might be a sign of simplicity as well. Suppose two computer programs having the same functionalities, but one having a less user-friendly interface. In that case, the one that is easier to use is also simpler, because it does not make us think.

In the next and final section we introduce two examples in which cognitive baroquism favors the emergence of chance-faking. The first example regards text messaging, whereas the second is about the so-called e-activism.

5 Simplicity and Consilience in Technology

Let us start with the example of text messaging. Before the advent of mobile phones appointments with friends, acquaintances, colleagues, etc., were arranged beforehand. And a delay may have caused great disappointment, since there was no way to inform about it. Mobile phones changed that. A text message allows a person to inform a friend he or she will be five minutes late. Texting may also serve to arrange a new place to meet in case of problems. Generally speaking, text messaging helps us deal with unexpected events in a unprecedented way comparing with the past. According to our definition of consilience applied to technology, mobile phone is certainly more consilient than the traditional (wired) phone, as it allows us to do what we had usually done with the traditional phone, plus it allows us to stay connected with our friends or acquaintance almost wherever we are.

However, we argue that in this case consilience can be reached at the expanse of simplicity. Mobile phone allows us to rely on much more expensive forms of communication with our friends and acquaintances. However, text messaging can mess things up, if misused. Indeed, when text messaging is used to inform our friend in case of an event totally independent from our will – a car crash, traffic jam, or whatever – our communication is just enhanced and enriched by a new and valuable chance, that is, flexible communication. A text may help avoid misunderstanding and possible complains due to one's delay. But, if one relies on such form of communication basically for avoiding to stick to the time (and maybe even the place) arranged for the appointment, then the chance provided by the mobile phone turns out to be counteractive, namely, a fake chance. Plus, it might create even worse misunderstanding than before. In a way one may argue that it would be much simpler to set an appointment for a time and stick to it, acting as if one did not have the chance to communicate with his or her friend.

At a more abstract level, the flexibility afforded by mobile communications takes away from the user an important communicative constraint people have usually made use of for planning and organizing their appointments. Such constraint is indeed a limitation, but it might also acquire a cognitive meaning insofar as it facilitates to keep communicating simple and therefore more reliable and just less ambiguous. Besides, the nature of such constraint – which *physically* makes mobile communication impossible – arranges one's cognitive niche in such a way that some affordances are immediately available whereas others are not. Conversely, the absence of physical constraints might literally intoxicate the cognitive niche with fake chances just like the case described above. For example, communication flexibility might afford a person to externalize her or his responsibility becoming a kind of *immoral* mediator [14]. Indeed, it is an immoral mediator from a particular ethical perspective according to which one should not adopt uncooperative behaviors and thus poison relationships with friends, acquaintances, colleagues, etc.

A similar yet different example is provided by the so-called e-activism. During the last decades a number of social applications has emerged potentially promoting grassroots activism, political inclusion, and an alternative conception of political activism based on the massive exploitation of information technologies. The social networking website Facebook is certainly the most successful among the social softwares and platforms for connecting people and creating consensus around a common cause. There are many examples that clearly demonstrated this potential. Consider recent campaigns, such as the international mobilization to prevent Sakineh, an Iranian woman accused of adultery and murder, from being stoned to death, or the anti-FARC campaigns reported in [17]: those campaigns are clearly devised by highly-committed and informed individuals, and then proposed unto mass public by means of mouth-to-ear dynamics (sometimes picked up and echoed by corporate news) and social networks. Astonishingly, about 100.000 people were said to confirm their participation within the Facebook group, demonstrating their solidarity with the Colombian citizens.

Sticking to the case of Facebook as representative, groups and fanpages offer the chance to get people involved in any political issue that one would like to bring up in the public sphere. Groups and fanpages allow 1) to make people's preferences visible and 2) to aggregate them so as to raise public concern about a certain issue (Sakineh, FARC, or whatever). There is no doubt that such tools are enhancing and empowering, if one looks at the chances that they potentially provide the users with. However, if one compares e-activism with traditional forms of activism, for instance, that of the 1960's, it seems that the former is characterized by superficiality and shallow connections resulting from desultory behavior [3,4]. As a matter of fact, social networks like Facebook drastically lower the bar for effective and actual political participation and involvement. However, just like in the case of text messaging, we claim that this simplification is just apparent: what actually happens is that one is just one click away from embarking on any kind of conceivable online campaign. Is that really enough for defining political participation and involvement?

In this case the technological tool plays the actual role of a bullshitting *para-democratic channel* where people are *de facto* "far removed from the citizenry and from the most direct exercise of democracy" [5]. This technological bullshitting effect is related to the fact the tool fakes the chance of democracy disregarding the *true* components of it, instead more or less intentionally flattering that kind of hallucinatory hyperindividualism started after the fall of the Berlin wall, when the new neoliberal era arose together with stock market aces taken off-guard by the dot-coms and other new agencies, then mortgages, and finally the deep current economic crisis triggered [5, p. 53]. A hyperindividualism paradoxically contradicted by the more and more prominent role of those business oriented informational macrostructures themselves, which instead at the end humiliate various aspects of the single citizens' dignity.

If it is true that it is much easier for a person to be exposed to many more chances to get politically involved, it is also true that such easiness may just promote superficial and degraded commitment due to the number of irrelevant requests one usually receives – a kind of information overloading. And so promoting actual fake "democratic" chances. Analogously to the case of text messaging, here again consilience is reached at the expense of simplicity. Traditionally, activism was considered as an effect of a sincere ideological commitment, and was usually focused within a well-defined field (i.e. women's rights, civil rights, environmental issues). Conversely, the emergence of new forms of activism mediated by social networking web sites like Facebook may make paradoxically harder political activism, if one goes beyond the "I click" or/and "I like" dimension which is characteristic to social tools like Facebook. In fact, the various requests for getting involved are rather chances for a desultory form of participation characterized by a digressive and shallow, random and occasional attitude.

6 Conclusions

In this paper we have illustrated that the activity of chance-seeking might be drastically impaired by the presence of fake chances resulting from bullshitting. We have pointed out that one of the main aspects of bullshitting – considered from a cognitive perspective – is the systematic dismissal of one of the most important pillars of scientific rationality: simplicity. What is important to stress here is that the bullshitter as a chance faker fails to take into account the importance of looking at what one can assume as true and what one cannot. In the last part of the paper, we have tried to show how this attitude might affect the way people use technology. More specifically, we have illustrated how a chance – potentially benefiting for cooperation and human relations – may turn out to be a fake chance due to the adoption of a bullshitting attitude.

References

1. Abe, A.: Cognitive chance discovery. In: Stephanidis, C. (ed.) UAHCI 2009, Part I. LNCS, vol. 5614, pp. 315–323. Springer, Heidelberg (2009)
2. Bardone, E.: Seeking Chances. From Biased Rationality to Distributed Cognition. Springer, Heidelberg (2011)
3. Bardone, E.: Unintended affordances as violent mediators: Maladaptive effects of technologically enriched human niches. The International Journal of Technoethics 2(4), 37–52 (2011)
4. Bertolotti, T.W., Bardone, E., Magnani, L.: Perverting activism. cyberactivism and its potential failures in enhancing democratic institutions. The International Journal of Technoethics 2(2), 14–29 (2011)
5. Brey, A., Innerarity, D., Mayos, G.: The Ignorance Society and Other Essays. Libros Infonomia, Barcelon (2009); translated from Spanish by M. Black
6. Cornella, A.: Com sobreviure a la infoxicació (how to survive infoxication) (2000), `http://www.infonomia.com/img/pdf/sobrevivir_infoxicacion.pdf`
7. Frankfurt, H.: On Bullshit. Princeton University Press, Princeton (2005)
8. Goffman, E.: Frame Analysis. An Essay on the Organization of Experience. Northeastern University Press, Boston (1986)
9. Hample, D., Warner, B., Young, D.: Framing and editing interpersonal arguments. Argumentation 23, 21–37 (2009)
10. Hample, D., Han, B., Payne, D.: The aggressiveness of playful arguments. Argumentation 24(4), 405–421 (2010)
11. Jackson, S.: Black box arguments. Argumentation 22, 437–446 (2008)
12. Magnani, L.: Chance discovery and the disembodiment of mind. In: Oehlmann, R., Abe, A., Ohsawa, Y. (eds.) Proceedings of the Workshop on Chance Discovery: from Data Interaction to Scenario Creation, International Conference on Machine Learning (ICML 2005), pp. 53–59 (2005)
13. Magnani, L.: Abductive Cognition. The Eco-Cognitive Dimension of Hypothetical Reasoning. Springer, Heidelberg (2009)
14. Magnani, L.: Understanding Violence. The Intertwining of Morality, Religion and Violence: A Philosophical Stance. Springer, Heidelberg (2011)

15. Magnani, L., Bardone, E.: Sharing representations and creating chances through cognitive niche construction. The role of affordances and abduction. In: Iwata, S., Oshawa, Y., Tsumoto, S., Zhong, N., Shi, Y., Magnani, L. (eds.) Communications and Discoveries from Multidisciplinary Data, pp. 3–40. Springer, Berlin (2008)
16. Misak, C.: Pragmatism and solidarity, bullshit, and other deformities of truth. Midwest Studies in Philosophy 32, 111–121 (2008)
17. Neumayer, C., Raffl, C.: Facebook for Global Protest: The Potential and Limits of Social Software for Grassroots Activism. In: Stillman, L., Johanson, G. (eds.) Proceedings of the 5th Prato Community Informatics & Development Informatics Conference (2008)
18. Odling-Smee, F.J., Laland, K.N., Feldman, M.W.: Niche Construction. The Neglected Process in Evolution. Princeton University Press, Princeton (2003)
19. Ohsawa, Y., McBurney, P. (eds.): Chance Discovery. Springer, Berlin (2003)
20. Peirce, C.S.: Collected Papers of Charles Sanders Peirce. Harvard University Press, Cambridge (1931-1958); Hartshorne, C., Weiss, P., (eds.), vol. 1-6, Burks, A. W., (ed.), vol. 7-8
21. Simon, H.A.: Altruism and economics. American Economic Review 83(2), 157–161 (1993)
22. Thagard, P.: The best explanation: Criteria for theory choice. The Journal of Philosophy 75(2), 76–92 (1978)
23. Thagard, P.: Conceptual Revolutions. Princeton University Press, New York (1992)
24. Thomas, H.J.: Are theories of imagery theories of imagination? An active perception approach to conscious mental content. Cognitive Science 23(2), 207–245 (1999)
25. Vrij, A.: Detecting Lies and Deceit Pitfalls and Opportunities. Wiley, New York (2008)

A Chance Favors a Prepared Mind: Chance Discovery from Cognitive Psychology

Hitoshi Terai and Kazuhisa Miwa

Graduate School of Information Science, Nagoya University,
Furou-chou, Chikusa-ku, Nagoya, Aichi, Japan
{terai,miwa}@is.nagoya-u.ac.jp
http://miwalab.cog.human.nagoya-u.ac.jp

Abstract. A chance has two contrary aspects: suddenness as an accidental finding and gradualness as the result of a prepared mind. Such duality of chance discovery resembles the insight process treated by problem solving researches. In this paper, we focus on the insight process in human problem solving, present a broad overview of its suddenness and the gradualness, and introduce our experimental results from the viewpoint of the duality of insight. We believe that our research findings will contribute to studies of chance discovery.

Keywords: aha experience, constraint relaxation, eye movement, impasse, incubation, insight, prepared mind, reconstruction, representation change.

1 Introduction

In the study of chance discovery, a chance is defined as an event or situation that may crucially affect decision-making or suggest that an important event will occur in the future. The studies of chance discovery address how we can exploit such chance to benefit the development of a new market, to avoid risk, etc.

Such aspects are not new ways of thinking about chance discovery. In data mining, methodologies that reveal the tendencies of past data and predict the future have been investigated for many years. The originality of the study of chance discovery is in the perspective of a chance. The word "chance" comes from the Latin *casus*, meaning luck. It means that one has the impression that a chance discovery arises from uncontrollable causes. Creators who noted important historical things often reported that their findings were blessings [4]. Such perspectives about chance discovery stress that there is no preparation to find a chance.

On the other hand, consider the proverb: "Chance favors a prepared mind." This perspective recognizes that chance is related to human factors. Where do these contradictions come from? Perhaps a prepared mind was unconsciously processed. Such duality of chance discovery, suddenness as an accidental finding

Y. Ohsawa & A. Abe (Eds.): Advances in Chance Discovery, SCI 423, pp. 33–48.
springerlink.com

and gradualness as a process of a prepared mind, resembles the insight process treated by problem solving researches.

In the paper, we focus on the insight process in the study of human problem solving, give a broad overview of the suddenness and the gradualness of insight, and introduce our experimental results from the viewpoint of the duality of insight.

2 Insight Problem Solving

In the Gestalt psychology conception, thinking is mainly classified as either reproductive or productive. Reproductive thinking is the application of previously acquired knowledge to a problem. On the other hand, productive thinking is the ability to go beyond experience and produce something new in response to the demands of a problem. Insight problem solving is categorized as the latter[10,26].

Insight problem solving is different from normal problem solving in many aspects[21]. In Gestalt psychology, the insight process is defined as follows. First, configuration of a problem and past experience prevent it from finding the relations required to solve a problem, creating an impasse for the problem solver. Finally, insight may arise through the reconstruction of a problem from a different perspective[16,10].

Box 1 shows an example of an insight problem that requires reconstruction of an incorrect recognition of the problem. As long as the problem solver treats it superficially, he or she cannot solve it. The problem's structure must be reconstructed to find the covert relations. In the problem, only the circle's radius is given; the diagonal's length cannot be obtained directly. However, the solution can be obtained through the following reconstruction. A rectangle has two diagonals. By focusing on the second diagonal, the problem solver realizes that it equals the rectangle (aha!). Thus, when the perspective of a problem is reconstructed, a solution is derived from that reconstruction result [8,11].

Box 1 Geometrical Problem

If the length of the radius is r, what is the length of line l?

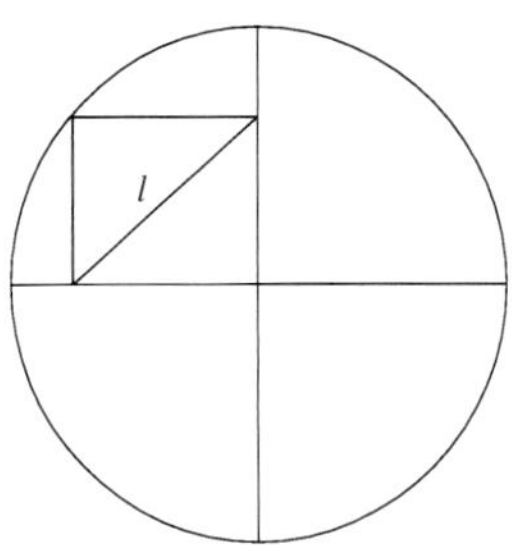

[8]

Six-year-old Karl Gauss solved a tedious arithmetic problem very quickly by reconstructing it[14]. Hoping to give himself a few minutes' peace, his teacher asked his class to add up the numbers 1 + 2 + 3 + 4 etc. up to 100. Hardly had the class begun to add up all the numbers when Gauss put his hand up with the answer: 5050. Gauss had reconstructed the problem. He made two numerical sequences: an ascending sequence from 1 to 100 and a descending sequence from 100 to 1. He could make a hundred 101 by adding the numbers at the same position on each sequence (1 + 100, 2 + 99, etc.). Adding the series 1 + 2 + 3 up to 100 is therefore the same as taking half of 10, 100; that is 5050. Thus, reconstruction occurs not only in perceptual relations but also in conceptual relations.

3 Duality of Insight Process

The moment of insight problem solving by reconstructing the problem seems to occur suddenly. On the other hand, a preparation phase has been identified during insight problem solving. In this section, we focus on such contradictory aspects, suddenness and gradualness, in the process of insight problem solving.

3.1 Suddenness of Insight

In the process of insight problem solving, a solution that seems to arise suddenly is generally known as an "aha! experience." Such suddenness, a key feature of insight problem solving, is also found in the recollections of historical scientific findings[25].

Metacognition in Insight Problem Solving. Metcalfe and Wiebe discussed the suddenness of insight problem solving from the viewpoint of metacognition by focusing on it both before and during insight problem solving[12]. In their experiment, participants were given both insight and non-insight problems and were required to rearrange them into a line going from the problem they thought they were most likely to solve to the one they were least likely to solve. The correlation results between the difficulty of the problems and the response time of each revealed that the participants fairly accurately predicted which non-insight problems they would be able to solve later, but they were unable to predict which insight problems they would solve.

Moreover, every 15 seconds while solving the problems, the participants were told to indicate their *feeling of warmth* (i.e., their perceived closeness to a solution). The result shows that the patterns of warmth ratings, which refract participant feelings about the approaching solution, differed for insight and non-insight problems. Non-insight problems showed a more incremental pattern than the insight problems. In general, the data indicate that non-insight problems were open to accurate predictions of performance, but the insight problems were opaque to such predictions.

Metcalfe and Wiebe's findings were as follows. The first is about metacognition prior to addressing insight problems. People can accurately predict the degree of difficulty of non-insight problems before solving them, but not in insight problem solving. The second is about metacognition in insight problem solving. People can accurately recognize a differential between the current state and the goal state in non-insight problems, but they cannot do so in insight problem solving; they feel that they are suddenly finding a solution.

Box 2 Box-Candle Problem

You have a candle, matches, and a box of tacks. Support the candle on the wall without dripping wax on the table.

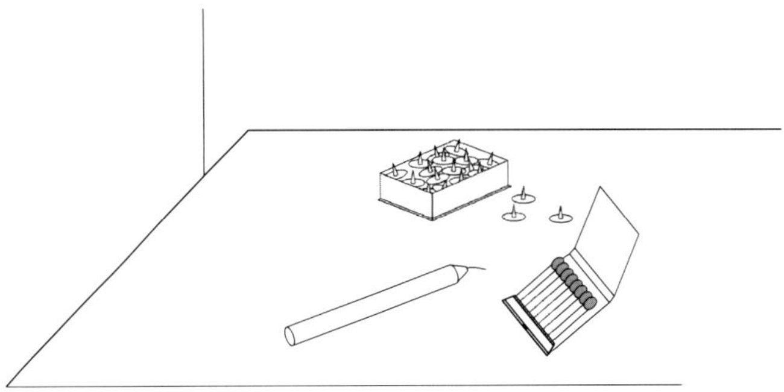

[3]

Representation Change. Furthermore, aspects of problems such as manner of formulation of problems change fundamentally across before and after finding a solution [11]. For example, in the box-candle problem (Box 2), a famous task used in studies of insight problem solving, subjects are given candles, matchsticks, and a box filled with tacks. The objective of the task is to attach a candle to the wall so that it will not drip wax on the table. Using the box filled with the tacks is essential to solve this problem. However, despite the fact that the solution of this problem is very simple, finding the solution is difficult because we embrace the illusion that the box filled with tacks must hold the tacks as a container (see Appendix for the solution). Such an assumption is automatically established; reconstructing the incorrect structure of problems is required to find solutions[10,16]. After establishing such reconstruction, if in the future the same problem is encountered, the impasse will not be met again.

These studies revealed that in insight problem solving, we cannot estimate current problem solving states and feel that representation suddenly changes with flash of illuminance at the moment when finding a solution. Such suddenness in the problem solving process is a crucial characteristic of insight problem solving.

3.2 Gradualness of Insight

In contrast to the sudden attainment of a solution described above, the process of insight problem solving develops through a prepared phase. For example, you experience an impasse, and you forget about the problem for a while, and then a solution suddenly appears while you are thinking about other things. Actually, many prominent creative individuals have reported similar introspective reports[25].

Incubation. Wallas outlined four major phases of information processing that may mediate innovative problem solving and creativity: (1) mental preparation, (2) incubation, (3) illumination, and (4) verification[25,17]. The initial preparation phase confronts an important problematic situation, conceptualizes the problem's core aspects, and makes exerted tentative unsuccessful attempts to reach a satisfactory resolution. Next, the incubation phase puts the problem aside and instead thinks about other matters for an extended period of time. Then, at some point during the incubation, there is a sudden shift to the illumination phase, where a penetrating flash of insight unexpectedly occurs about an appropriate satisfying resolution to the original problematic situation. Given the attained insight, the final verification phase concentrates on the details of the resolution or determines that it was applied successfully.

Next, we introduce two major studies about the effect of the incubation phase during solving insight problems. Silveira showed its effect by conducting psychological experiments[18,17] in which the participants were given the necklace problem shown in Box 3 (see appendix for the solution). They were instructed to put the problem aside as an incubation phase during problem solving; they resumed solving the problem after the incubation phase. The interruption time was controlled as an experimental factor.

The experiment results showed that 64% and 85% of the participants in the experimental conditions with short (30 min) and long (4 h) incubation phases found the solution, whereas only 55% of the participants in the control condition found the solution without the incubation phase.

Smith and Blankenship examined the effect of fixation and incubation in the insight process using remote associate tests (RAT)[20]. Box 4 shows an example of a RAT problem (see appendix). In the experiments, there were two experimental factors: fixation (fixation/nonfixation), and incubation (incubation/no-incubation). For RAT problems, the participants must think of a word that in combination with each of the three test words makes a two-word phrase or a compound word. For example, the solution to the RAT problem "TYPE / GHOST / STORY" is "WRITER." To lead the participants to the initial fixation, words were given as misleading clues (in this example, "STYLE / GOBLIN / TALE" were given at the same time).

The results of the experiments revealed that incubation significantly facilitated problem solving in the fixated group, but there was no significant effect in the non-fixated group.

Incubation unrelated to the confronted problems affected improvements to the problem solving and indicate that there is a preparation process that is different from conscious effort; such a process has a significant role in problem solving. In the next section, we discuss this unconscious process of a prepared mind from the viewpoint of the relaxation of mental constraints.

Box 3 Necklace Problem

You have four separate pieces of chain that are each three links long, as shown on the left side of the figure. It costs 2 cents to open a link and 3 cents to close a link. All links are closed at the beginning of the problem. Your goal is to connect all 24 links into a single circle, as shown on the right side of the figure. The total cost must not exceed 15 cents.

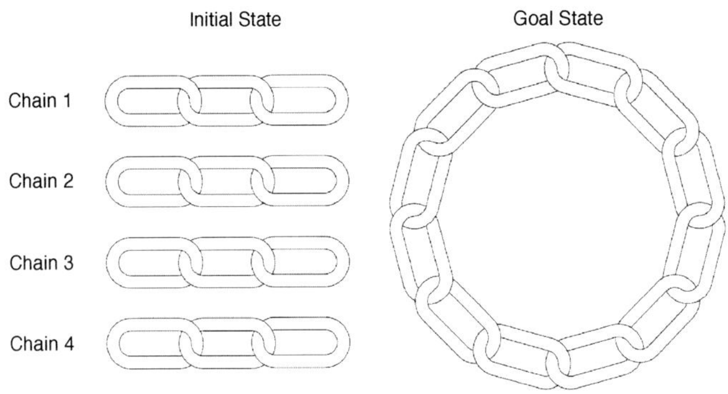

[18]

Box 4 Remote Associate Test

Find a single word that forms a common word or phrase with each of the three words in the upper row.

TYPE	GHOST	STORY
style	goblin	tale

[20]

Mental Constraint Relaxation. In some cases the structure of problems and previous knowledge act as negative constraints and inhibit problem solving. Therefore problem solvers may encounter an impasse around which they can't proceed. Box 5 shows the nine dots problem, which is one of the most famous experimental tasks in the study of insight problem solving (appendix).

Even though this problem is seemingly very simple, many people cannot solve it. The difficulty arises from an unconscious, incorrect assumption that the lines should begin and end on dots and were drawn within a square made by eight dots outside of the nine dots[9,16,15,2].

Box 5 Nine Dots Problem

Connect all nine dots using four straight lines without lifting your pen.

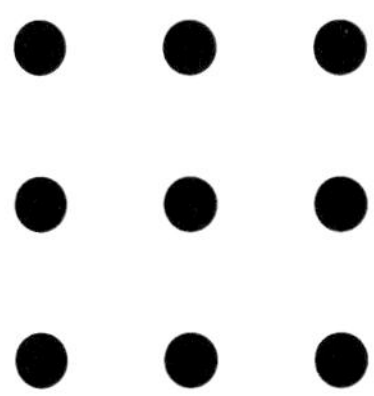

[27]

Recently, the insight process has been described from the viewpoint of mental constraint relaxation. First, mental constraints arise from past experiences and/or the structures of problems, and the problem solver meets an impasse because these constraints prevent a solution from being reached. Such mental constraints mean the false formulation of problem structure and incorrect interpretation of problems. Through problem solving, these mental constraints are gradually relaxed. A search that does not follow these mental constraints gradually increases, and then the problem solver reaches a solution[6,13].

Knoblich et al. analyzed mental constraints using the matchstick arithmetic problem that consists of a false arithmetic statement written with roman numerals (I, II, III, etc.), arithmetic operation (+, -), and equal signs constructed out of matchsticks[7]. Box 6 shows an example that was presented to the participants in their study. The goal is to move a single matchstick so that the initial false statement is transformed into a true arithmetic statement (appendix).

In their study, they argued that impasses are broken by changing the problem representation and described two hypothetical mechanisms for representational change: the relaxation of the constraints of the solution and the decomposition of perceptual chunks.

Problem solvers must relax one or more of three constraints: value, operator, and tautology. When mapped onto three representational levels, the three constraints each have different scopes. When the value constraint is relaxed, the problem solver can change a numerical value to another (e.g., from "IV" to "VI"). When the operator constraint is relaxed, the problem solve can change an arithmetic function to another (e.g., from "+" to "="). When the tautology constraint is relaxed, the problem solver can change an arithmetic statement to a tautological statement (e.g., from "III = III + III" to "III = III = III").

On the other hand, familiarity with a class of objects or events creates patterns that capture the recurring constellations of features or components. Knoblich et al. proposed that the probability that a chunk will be decomposed is inversely proportional to its tightness. For example, the problem solver can decompose "VI" to "V" and "I" more easily than "V" to a pair of "I"s because the former's chunk is loose, but the latter's is tight. They concluded that a constraint with a wide scope is more difficult to relax than one with a narrow scope. A tight chunk is more difficult to decompose than a loose one, and constraints and chunks independently lead the problem solver to an impasse.

Box 6 Matchstick Arithmetic Problem

Transform the false equation into a true equation by moving a match.

[7]

4 Experiment

As described above, previous researches either highlighted the suddenness or the gradualness of the nature of the insight process. However, two contrary natures of the insight process should be understood as its bilateral character. Until insight is consciously perceived as a sudden event, the preparing process should proceed unconsciously. However little research has addressed such duality of insight.

In this section, we introduce our study that revealed the duality of the insight problem solving.

4.1 Dual Space Search

We investigated the process of insight problem solving using a discovery task. The discovery process is the interaction between searching behavior for hypotheses held by the problem solver and instances in the environment used to construct and test them. Simon and Lea discussed such a discovery process as the interaction between searching for a rule space and an instance space[19]. Such an aspect has been established as a dual space search model of the scientific discovery process by conducting many psychological experiments and constructing computational models[5].

In our study, we conducted a psychological experiment to capture the duality of the insight process based on the dual space search model. To analyze the insight process as a dual space search, searching for a mental problem space is defined as searching for hypothesis space, and gathering data during hypothesis formation and testing is defined as searching for data space.

Verbal Protocole. In our task, since the participants reported their hypotheses, we captured the process of their hypothesis searches as verbal reports.

Eye movement. We also captured their searching for data as eye movements.

4.2 Method

Subjects. Twenty-four undergraduate students participated in this experiment.

Task. In this study, we proposed and used a discovery task that asked subjects to find a rule for predicting a digit. In the task, subjects must find a target rule as a solution through hypothesis formation and testing.

An example screenshot of the task is shown in Fig. 1. The display consists of three slots, and in each slot a single digit rotates at a speed that prevents the subjects from perceiving each digit. A history data window indicates the past four trials. The digit in the third slot is controlled by an unknown rule (target rule). If subjects find the target rule, then they can predict the digit in the third slot. The mission of the subjects is to find the target rule and predict the digits in the third slot.

Subjects are required to predict the digit in the third slot after the two digits in the first and second slots stop rotating. A series of procedures, which includes stopping the first and second slots and then predicting and confirming the third digit, is called a trial; trials are repeated until subjects find the target rule. A history data window below the three slots shows the results of the past four trials.

The rules reported in each trial by subjects are called hypotheses, which are proposed in the process of hypothesis formation and testing. This is a discovery task that requires insight. Therefore, the task is manipulated to lead the subjects to find a sham rule called a blocking hypothesis, which differs from the target rule. The target rule and blocking hypothesis are shown below. The target rule states: "the third digit in the nth trial is determined by adding three to the third digit in the n-1th (previous) trial" with a vertical relation on the display. The

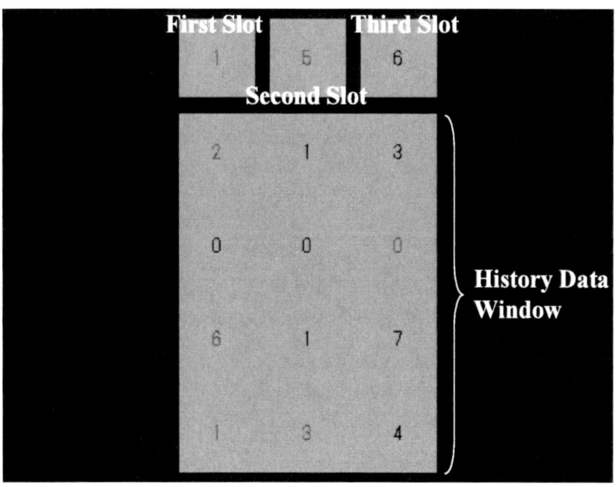

Fig. 1. Example screenshot of discovery task used in this study

blocking hypothesis states: “the third digit is equal to the sum of the first and second digits” with a horizontal relation.

Subjects are required to predict the digit in the third slot after the two digits in the first and second slots stop rotating. In the initial eight trials, by controlling the first and second digits, the third digit is consistent with the sum of the first and second digits while maintaining the target rule across the vertical row (see example display in Fig. 1; this rule is confirmed as follows: $1 + 3 = 4$, $6 + 1 = 7$, $0 + 0 = 0$, $2 + 1 = 3$, $1 + 5 = 6$ in each column as a blocking hypothesis, and a series of 4, 7, 0, 3, 6 from bottom to top in the third row as the target rule). Therefore, subjects are guided to form a blocking hypothesis with the horizontal relation as a sham rule. After the first eight trials, a digit predicted by the blocking hypothesis gradually disagrees with an actual third digit. That is, the blocking hypothesis doesn't predict the third digits but the target rule is still maintained. Consequently, from the ninth trial subjects begin to receive negative instances called negative feedback that disconfirms the blocking hypothesis.

Definition of Types of Hypothesis Spaces. The process through which subjects discover the target rule is considered a process of searching for a hypothesis space. A group of hypotheses that share a common regularity constitutes a hypothesis space; therefore the target rule and the blocking hypothesis belong to different problem spaces. In this study, the relation between hypotheses and hypothesis spaces is defined as follows:

Blocking Hypothesis Space: A set of hypotheses characterized by a horizontal relation is defined as a blocking hypothesis space.
Blocking Hypothesis: described above.
Horizontal Hypotheses: rules characterized by a horizontal relation other than the blocking hypothesis (e.g., “ the third digit is equal to the subtraction of the second digit from the first digit”).
Target Space: A set of hypotheses characterized by a vertical relation is defined as a target space.
Target Rule: described above.
Vertical Hypotheses: rules characterized by a vertical relation other than the target rule (e.g., “the third digit is equal to the sum of the third digits in all past trials in the history data window”).

In addition to these hypotheses, subjects reported hypotheses characterized by both horizontal and vertical relations, which are involved neither in the blocking hypothesis space nor in the target space, such as identical digits arranged diagonally.

The search for each hypothesis space described above is demonstrated in Fig. 2. Running into an impasse and finding a target rule are shown as follows when the insight process is considered a search for problem spaces. An impasse means that after finding a blocking hypothesis, subjects are fixated on searching for the blocking hypothesis space that consists of similar hypotheses with the horizontal relation to the blocking hypothesis, even if the blocking hypothesis is disconfirmed. It is necessary to exit from the search for the blocking hypothesis space

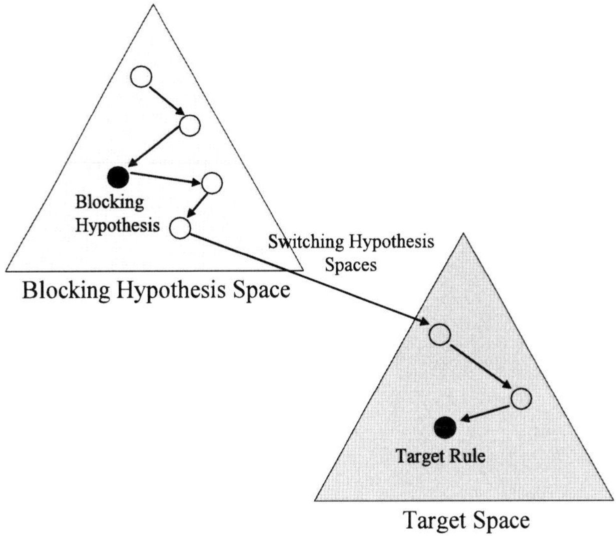

Fig. 2. Relation between Blocking Hypothesis and Target Rule

and move to a search for the target space to find the target rule. A previous study indicated that once the blocking hypothesis was formed, many subjects could not extricate themselves from incorrect constraints and failed to find the target rule even though it is very simple.

Procedure. Subjects start a trial, report a predicted rule as a hypothesis after the first and second slots stopped, and discontinue the third slot to confirm the hypothesis. Such trials consisting of hypothesis proposing and testing lasted for a maximum of 55 minutes until finding the target rule. If subjects reported the target rule continuously five times, then it was considered that they had found the target rule.

4.3 Results

We excluded eight of the twenty-four subjects from analysis because they could not form a blocking hypothesis during the initial eight trials or could not provide any fine eye movement data. Five of the sixteen subjects found the target rule and were categorized as the successful group; eleven subjects did not find the target rule and were categorized as the unsuccessful group. In this paper, we are only concerned with the results of the successful group (for details, see [23]).

Transition of Hypothesis Space Search. The proposed hypotheses were categorized into one of the five types defined in section 4.2; how they changed was also analyzed. In this experiment, we did not provide a second evaluater because in each trial the subjects were required to verbally report proposed hypotheses, which we clearly captured. Fig. 3 shows the transition of the proposed hypotheses in the successful group. The horizontal axis indicates the number of trials, and the vertical axis indicates each type of hypothesis and the hypothesis space described above. On the vertical axis in Fig. 3, the blocking hypothesis and hypotheses with horizontal relations correspond to hypotheses within the blocking hypothesis space in Fig. 2, and the target rule and hypotheses with vertical relations correspond to hypotheses within the target space in Fig. 2.

Fig. 3 shows that all successful subjects found the blocking hypothesis by the time they reached the ninth trial. After the ninth trial, they began to receive negative instances for the blocking hypothesis and to propose other hypotheses than the blocking hypothesis. However, hypotheses outside of the blocking hypothesis space were almost never proposed, confirming that the subjects continued to search for the blocking hypothesis space. This result indicates that the subjects had run into an impasse. Fig. 3 also shows that discovering the target rule suddenly seemed to occur while escaping from a state where subjects searched for the blocking hypothesis space, rather than through gradual shifting of potential hypothesis spaces. For example, subject I5 suddenly found the target rule from the fourteenth to fifteenth trials with an abrupt shift from searching for the blocking hypothesis space to the target space. Correspondingly, the other successful subjects also found the target rule suddenly after abandoning their search for the blocking hypothesis space.

Transition of Data Space Search. Next, we analyze the process of subjects searching for data spaces that led to finding the target rule using eye movement data, obtained as transition patterns of the fixation of eye movement. In this study, eye movements from nth to n+1th fixation points were categorized as (1) horizontal, (2) vertical, (3) diagonal, or (4) fixed.

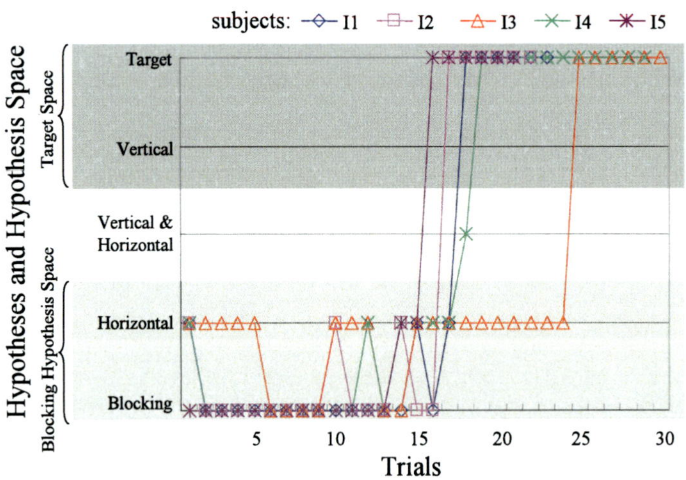

Fig. 3. Transition of hypothesis search in successful groups

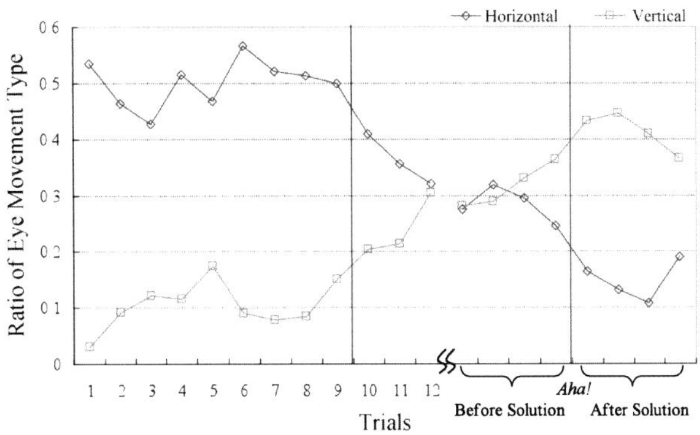

Fig. 4. Transition of data search in successful groups

The transition of the horizontal and vertical eye movements of successful subjects is shown in Fig. 4. The horizontal axis shows the number of trials: the first nine trials, three trials after negative feedback was given, and four trials before and after the target rule was discovered. The vertical axis indicates the ratio of each type of eye movement to all types of eye movement (horizontal, vertical, diagonal, and fixed). Fig. 4 shows that the horizontal eye movement, corresponding to searching for the data space correspond to the blocking hypothesis space, dominated until the subjects reached the ninth trial. By contrast, after they were given negative feedback, the ratio of horizontal eye movement gradually decreased but the ratio of vertical eye movement gradually increased. This result indicates that the search for the data space correspond to the blocking hypothesis space gradually decreased and the search for the data space correspond to the target space gradually increased. The gradual shift of searching for data spaces was captured in subject eye movement data.

4.4 Summary

In psychological experiments, we captured the transitions of hypothesis space search and data space search in the process of hypothesis formation and testing using subjects' verbal reports and eye movement analysis. From the viewpoint of the former, regarding their hypotheses as hypothesis space search, discovering the target rule seemed to occur suddenly following an impasse. From the viewpoint of the transition of the proposed hypotheses the experimental result showed exactly that in insight problem solving, a solution is found with flash of a illuminance called the "aha experience" that characterizes the suddenness of the insight process[1,12]. On the other hand, the analysis of subject eye movement as data space search revealed that searching for data gradually varied with the development of problem solving after negative feedback was given.

5 Conclusion

In this paper, we argued that in insight problem solving tasks, a prepared mind precedes the "aha experience". Such a process of reaching insight resembles the requirements of chance discovery. How do you help and lead a prepared mind to chance discoveries? Many challenges remain. To answer these questions, we must develop interdisciplinary studies between cognitive researches about understanding what is chance for humans and information engineering about specific techniques that encourage chance discovering.

Moreover, we introduced our research that attempted to reveal the prepared process of insight problem solving using eye movement data. A more direct way to capture unconscious processes is measuring brain activity. In recent years, such non-invasive brain imaging apparatus as Near InfraRed Spectroscopy (NIRS) and functional Magnetic Resonance Imaging (fMRI) have been developed. Future developments are expected to study the effective use of these measurement techniques. In addition, another approach is to construct a computational model that may represent the interaction of conscious and unconscious processes to explain the duality of the insight process. In problem solving research, there are already related studies (e.g., [22]). We have already begun a study based on a computational model approach[24].

References

1. Davidson, J.E.: The suddenness of insight. In: Sternberg, R.J., Davidson, J.E. (eds.) The Nature of Insight. ch. 4, pp. 125–155. MIT Press, Cambridge (1995)
2. Dominowski, R.L., Dallob, P.: Insight and problem solving. In: Sternberg, R.J., Davidson, J.E. (eds.) The Nature of Insight. ch. 2, pp. 33–62. MIT Press, Cambridge (1995)
3. Duncker, K.: On problem-solving. Psychological Monographs 58(270), 1–113 (1945)
4. Ferrucci, P.: Inevitable Grace: Breakthroughs in the Lives of Great Men and Women: Guides to Your Self-Realization. St Martins Pr. (1990)
5. Klahr, D., Dunbar, K.: Dual space search during scientific reasoning. Cognitive Psychology 12, 1–55 (1988)
6. Knoblich, G., Ohlsson, S., Raney, G.E.: An eye movement study of insight problem solving. Memory & Cognition 29(7), 1000–1009 (2001)
7. Knoblich, G., Ohlsson, S., Haider, H., Rhenius, D.: Constraint relaxation and chunk decomposition in insight problem solving. Journal of Experimental Psychology: Learning, Memory, and Cognition 25(6), 1534–1555 (1999)
8. Kohler, W.: The task of Gestalt psychology. Princeton University Press, Princeton (1969)
9. Maier, N.R.F.: Reasoning in humans: I. on direction. Journal of Comparative Psychology 10, 115–143 (1930)
10. Mayer, R.E.: Gestalt -thinking as restructuring problems-. In: Thinking, Problem Solving, Cognition, 2nd edn. ch. 3, pp. 39–78. W.H. Freeman and Co., New York (1992)
11. Mayer, R.E.: The search for insight: Grappling with gestalt psychology's unanswerd questions. In: Sternberg, R.J., Davidson, J.E. (eds.) The Nature of Insight. ch. 1, pp. 3–32. MIT Press, Cambridge (1995)

12. Metcalfe, J., Wiebe, D.: Intuition in insight and noninsight problem solving. Memory & Cognition 15(3), 238–246 (1987)
13. Ohlsson, S.: Information-processing explanations of insight and related phenomena. In: Keane, M., Gilhooley, K. (eds.) Advances in the Psychology of Thinking. ch. 1, pp. 1–44. Harvester Wheatsheaf, London (1992)
14. Robertson, S.I.: Problem solving. Taylor & Francis (2001)
15. Rumelhart, D.: Introduction to human information processing. Wiley, New York (1977)
16. Scheerer, M.: Problem solving. Scientific American 208, 118–128 (1963)
17. Seifert, C.M., Meyer, D.E., Davidson, N., Patalano, A.L., Yaniv, I.: Demystification of cognitive insight: Opportunistic assimilation and the prepared-mind perspective. In: Sternberg, R.J., Davidson, J.E. (eds.) The Nature of Insight. ch. 3, pp. 65–124. MIT Press, Cambridge (1995)
18. Silveira, J.M.: Incubation: The effect of interruption timing and length on problem solution and quality of problem processing. Unpublished doctoral dissertation, University of Oregon, Eugene (1971)
19. Simon, H.A., Lea, G.: Problem solving and rule induction: A unified view. In: Gregg, L. (ed.) Knowledge and Cognition, Erlbaum, Hillsdale (1974)
20. Smith, S.M., Blankenship, S.E.: Incubation and the persistence of fixation in problem solving. American Journal of Psychology 104, 61–87 (1991)
21. Sternberg, R.J., Davidson, J.E. (eds.): The Nature of Insight. MIT Press, Cambridge (1995)
22. Sun, R.: Duality of the Mind. Lawrence Erlbaum Associates, Publishers, London (2002)
23. Terai, H., Miwa, K.: Insight problem solving from the viewpoint of constraint relaxation using eye movement analysis. In: Proceedings of the 4th International Conference of Cognitive Science, pp. 671–676 (2003)
24. Terai, H., Miwa, K.: Sudden and gradual processes of insight problem solving: Investigation by combination of experiments and simulations. In: Proceedings of 28th Annual Meeting of the Cognitive Science Society, pp. 834–839 (2006)
25. Wallas, G.: The art of thought. Harcourt Brace Jovanovich, New York (1926)
26. Weisberg, R.W.: Prolegomena to theories of insight in problem solving: A taxonomy of problems. In: Sternberg, R.J., Davidson, J.E. (eds.) The Nature of Insight. ch. 5, pp. 157–196. MIT Press, Cambridge (1995)
27. Wickelgren, W.A.: How to solve problems. Freeman, Sanfracisco (1974)

Appendix: Solution of the Insight Tasks

Candle-Box Problem

Solution: Empty the tackbox, and tack it to the wall. Set the candle on the platform formed by the box.

Four Chain Problem

Solution: Separate one of the four chains to three rings (cost: 6 cents). Connect the remaining three chains to each other by using three rings (cost: 9 cents).

Remote Associate Test

Solution: *writer* (type*writer*, ghost*writer*, story*writer*)

Nine Dots Problem

Solution: See Fig. 5.

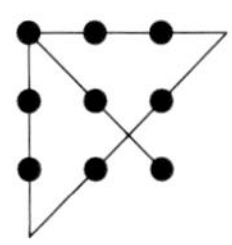

Fig. 5. Solution of nine dots problem

Match Stick Problem

Solution: See Fig. 6. Decompose IV to I and V, and move I from the left side of V to the right side.

Fig. 6. Solution of match stick problem

Chance Discovery with Self-Organizing Maps: Discovering Imbalances in Financial Networks

Peter Sarlin

Turku Centre for Computer Science – TUCS, Department of Information Technologies, Åbo Akademi University, Joukahaisenkatu 3-5, 20520 Turku, Finland
Peter.Sarlin@abo.fi

Abstract. In this chapter, we introduce the Self-Organizing Map (SOM) from the viewpoint of Chance Discovery. The SOM paradigm supports several principal parts of Chance Discovery: visualization of temporal multivariate data, discovering rare clusters bridging frequent ones, detecting the degree of event rarity or outliers, and dealing with continuously evolving structures of real world data. Here, we further enhance the standard SOM paradigm by combining it with network analysis. Thus, we enable a simultaneous view of the data topology of the SOM and a network topology of relationships between objects on the SOM. The usefulness of the Self-Organizing Network Map (SONM) for Chance Discovery is demonstrated on a dataset of macro-financial measures. While the standard SOM visualizes country-specific vulnerabilities by positions on the map, the SONM also includes bilateral financial exposures to show the size of linkages between economies and chances of shock propagation from one country to the rest of the world.

Keywords: Chance Discovery, Self-Organizing Map (SOM), network analysis, exploratory data analysis, financial stability.

1 Introduction

Since the turn of the 21st century, a large set of tools have been developed for Chance Discovery – where *chances* are rare, significant opportunities or risks of the future and *discovery* is the search for information with significant impact on human decision making [1]. In this sense, chance discovery differs from conventional search for tendencies in datasets (e.g. knowledge discovery or discovery science) by exploring possibilities of rare, novel and exceptional occurrences with significant effects on the system rather than solely those based upon historical trends. While some research focuses on automated machine discovery of chances [e.g. 2], a more common convention is to let, or preferably enable, human beings discover chances. Thus, machine automation is oftentimes combined with human interaction for discovery through human-computer interaction in a collaborative environment [3]. Common interactive visualization tools for discovering candidate chances, such as events linking multiple clusters of co-related frequent events, are the C4.5 decision-tree algorithm [4], Correspondence Analysis [5] and KeyGraph [6]. The interface used in [7] enables

Y. Ohsawa & A. Abe (Eds.): Advances in Chance Discovery, SCI 423, pp. 49–61.
springerlink.com

assessing data relationships of the C4.5 decision tree, KeyGraph is an algorithm for mapping co-occurrence of terms from textual data, and Correspondence Analysis projects categorical data onto a two-dimensional plane as principal component analysis does with continuous data. A self-evident next step is the application of non-linear projection techniques, such as the Self-Organizing Map (SOM), for assessing the data topology and frequency distribution of occurrences and events.

The SOM, proposed by Kohonen [8–9], is a widely used general-purpose tool for abstraction of multivariate data. It is an unsupervised and non-parametric neural network approach that pursues data reduction via clustering and dimensionality reduction via projection. While the SOM clustering can be thought of as a spatially constrained form of k-means clustering, the SOM projection is a mapping into a low-dimensional space maintaining the neighborhood relations in data. Further, the SOM projection differs from other projection methods, such as Multidimensional Scaling, by mapping data onto a lattice of nodes rather than into a continuous space. In the early days of the SOM, information extraction was mainly facilitated by visual analysis of a U-matrix, where a color code between all neighboring nodes indicates their average distance [10]. Lately, the SOM has, however, not only been utilized as a stand-alone clustering technique, its output has also been used as input for a second stage of two-level clustering [11]. In this chapter, we facilitate information extraction by applying Ward's hierarchical clustering [12] for finding second-level clusters of SOM nodes.

As the SOM enables an easily interpretable format of complex data, it is a common tool for exploratory data analysis, and thus also has potential for Chance Discovery. Indeed, since the SOM is not projecting into a continuous space, as e.g. Multidimensional Scaling, the data topology of the SOM can be related to the network topology of for example KeyGraph. In KeyGraph, one type of chance is represented by events bridging multiple clusters of strongly co-related frequent events. Similarly, as the SOM attempts to maintain neighborhood relations in data, not absolute distances, rare nodes (clusters) neighboring (in terms of Euclidean distance) frequent ones can be regarded as candidate chances which have been rare in the past but may have a high impact in the future. The SOM also presents means to deal with continuously evolving structures of real world systems, a common obstacle in the fields of prediction and forecasting that has also been stressed in Chance Discovery [13]. By locating data representing a currently occurring high-impact event, being it in a different location than historical trends point to or not, the SOM provides information of data in their neighborhood where similar high-impact events may occur. Thus, even though longer historical trends would point to other locations on the map, as the system changes the SOM sheds light on recent crucial areas in the data space.

In this chapter, usefulness for Chance Discovery is further enhanced by, to the best of our knowledge, first time in the literature combining the basic SOM paradigm with network analysis. While the standard SOM paradigm provides a, e.g., two-dimensional mapping of the inputs, it does not account for simultaneous relationships or links in data. Individually, visual clustering of social networks has been directly performed by applying the SOM on links or relationships in data [14]. However, we extend the basic SOM with a superimposed KeyGraph-style network analyzer for assessing data dependencies. The Self-Organizing Network Map (SONM) relates to Chance

Discovery by enabling visual highlighting of risks and opportunities in significant high-impact data points on an ordered mapping. In particular, the systemic relevance relates to, for instance, the analysis of bankruptcy contagion with a time-ordered KeyGraph [15]. In this sense, we combine the data topology of the SOM with a network topology of data relationships.

We demonstrate the usefulness of the SONM for Chance Discovery on a dataset of macro-financial measures. The SONM takes two types of inputs: features and relationships. The feature inputs for the SONM are standard indicators of macro-financial vulnerabilities of 28 advanced and emerging market countries spanning 1990–2010 used by Sarlin and Peltonen [16]. The inputs for the network analysis are financial linkages between the 28 countries represented by a matrix of bilateral asset exposures. While the macroprudential indicators show country-specific vulnerabilities by positions on the SOM, the financial exposures show the size of linkages and potential shock propagation to the rest of the world. This enables analyzing the spread of financial instabilities from two points of views: the data topology indicates propagation of financial distress to similar macro-financial conditions, while the network topology indicates shock propagation through asset-based financial linkages.

The chapter is structured as follows. Section 2 discusses the SOM, network analysis and the SONM. In Section 3, the SONM is applied on macro-financial indicators. Section 4 concludes by presenting our key findings and future research directions.

2 Methodology

2.1 Self-Organizing Maps

The model in this chapter is trained with the package Viscovery SOMine for its visualization and interaction capabilities. The implementation is based upon the Kohonen's [8–9] standard Euclidean SOM, but applies a combination of techniques and heuristics for speeding up the computations, such as linear initialization, batch algorithm, growing SOMs and short-cut winner search. These deviations from the original SOM are commonly used and well-documented in the literature (see Kohonen's comprehensive monograph [9] for details and reviews). The SOM grid consists of a predefined number of nodes m_i (where i=1,2,…,M), which represent reference vectors of the same dimensionality (number of inputs) as the actual data set. Generally, the implemented SOM-based algorithm operates according to five steps:

1. Pre-process the columns of input data x_j such that they have a desired weighting in the SOM, e.g. equal weighting.
2. Compare all data points x_j with all nodes m_i to find for each data point the nearest node m_c in terms of Euclidean distance (i.e., best-matching unit, BMU)

3. Update each node m_i to averages of the attracted data, including data located in a specified neighborhood σ
4. Repeat steps 2 and 3 a specified number of times
5. Group nodes into a reduced number of clusters of nodes using Ward's hierarchical clustering [12].

The parameters of the SOM are the following: number of nodes M, map format (ratio of X and Y dimensions), neighborhood radius σ, and number of training iterations t. Large radii result in stiff maps that stress topology preservation at the cost of quantization accuracy, while small radii lead to a standard k-means clustering with no topology preserving projection. The number of nodes sets the granularity of the SOM, which affects its generalization capability and quantization error.

For an easily interpretable but still descriptive model, we use a two-dimensional grid as the output of the SOM. Here, we use a simplistic black and white two-dimensional grid as we want to facilitate interpretation of the superimposed network model. The multidimensionality of the SOM grid is in this chapter visualized by feature planes. For each individual input, a feature plane represents the distribution of its values on the two-dimensional grid, where dark to light shades represent high to low values according to a color scale below each feature plane. Each unique point represents the same node on all feature planes, as they are different views of the same grid.

2.2 Network Analysis

Network analysis, or link analysis, can be seen as the exploration of crucial relationships and associations between a large set of objects that may not be apparent from assessing isolated data. The network of relationships is most often expressed in matrix form, where the link between object k and l in matrix $\boldsymbol{A}$ is represented by element $\boldsymbol{A}_{kl}$. The size of the matrix equals $n*n$, where n is the number of objects. In this sense, the matrix of a directed graph can be read in two directions: rows of $\boldsymbol{A}$ represent the relationship of object k to object l and columns of $\boldsymbol{A}$ represent the relationship of object l to object k. Obviously, an undirected graph differs by only having one directionally independent relationship between the objects k and l.

In Fig. 1, we exemplify the visualization of a standard directed network on a dataset of bilateral financial exposures between countries, which is later used as an input to the Self-Organizing Network Map (SONM). The figure shows a graph of financial relationships between countries where objects are represented by nodes and bilateral relationships by edges. Number of objects n equals 16, giving us a matrix $\boldsymbol{A}$ of the form 16x16, where each element $\boldsymbol{A}_{kl}$ represents the size of financial linkages between country k and l. Node size of each country is scaled based on the sum of exposures to other countries. The thickness of each edge represents the size of exposure to total exposures of each country, where the color of the edge indicates the address of the exposure holder.

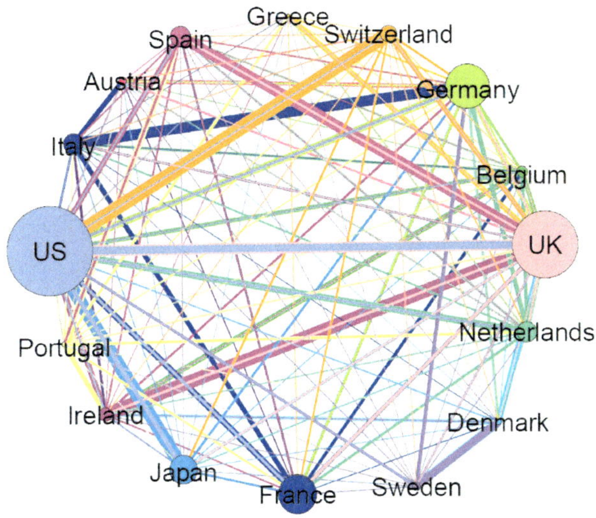

Fig. 1. A network of bilateral financial linkages for G3 countries in 2010, where nodes represent countries and edges linkages

2.3 Self-Organizing Network Map (SONM)

The Self-Organizing Network Map (SONM) combines the above methodologies, the SOM and network analysis, for simultaneous assessment of both the data and network topologies. The combinatorial approach superimposes network relations on top of a standard SOM by visualizing relationships between objects on the SOM grid. Thus, a SOM model is first trained using the above stated standard SOM methodology with data vectors x as its input. Then, nodes and labels of objects under analysis, say objects k and l, are projected to their BMUs of the SOM grid based upon their data $x_{j=k}$ and $x_{j=l}$. After that, relations between objects k and l are visualized by creating edges between the locations of the BMUs of x_k and x_l on the two-dimensional SOM using elements $\boldsymbol{A}_{kl}$ and $\boldsymbol{A}_{lk}$ of the network matrix $\boldsymbol{A}$. Thus, we visualize simultaneously the data topology of the SOM and a network topology of pure data relationships. While these two topologies have thusfar been mainly assessed in isolation, they are oftentimes highly interrelated as changes in one of the topologies may have significant implications on the stability of the combined topology.

3 Discovering Financial Imbalances

3.1 The Self-Organizing Map for Monitoring Financial Stability

This chapter replicates the Self-Organizing Financial Stability Map developed by Sarlin and Peltonen [16]. They put forward a robust framework for computing and evaluating the SOM for visual financial stability surveillance. The model uses the

SOM for mapping economies on a two-dimensional plane based upon financial vulnerabilities and for disentangling the individual sources of vulnerabilities impacting on systemic risks. It uses as inputs a set of vulnerability indicators commonly used in the macroprudential literature (see e.g. [17–19]) and class information representing pre-crisis, crisis, post-crisis and tranquil periods. The quarterly dataset consists of 28 countries, of which 10 are advanced and 18 emerging economies, spanning from 1990:1–2010:3. More precisely, we are dealing with a panel dataset consisting of both a cross-sectional and a temporal dimension. To control for cross-country differences and fixed effects, each column of the input data is transformed into its historic country-specific percentile distribution.

As is of central importance in Chance Discovery, such as in [2], we account for the rarity of high-impact events as well as asymmetric costs of false signals of crisis and tranquil periods when evaluating the model. The parameters and thresholds of the SOM are chosen as to minimize a loss function of a policymaker, where utility depends on his relative preferences between false negatives and false positives. The quality of the model is tested by predicting out-of-sample the global financial crisis of 2007–2009, by comparing its performance with a conventional statistical technique and by a number of robustness tests. The two-dimensional SOM can be found in Fig. 2.

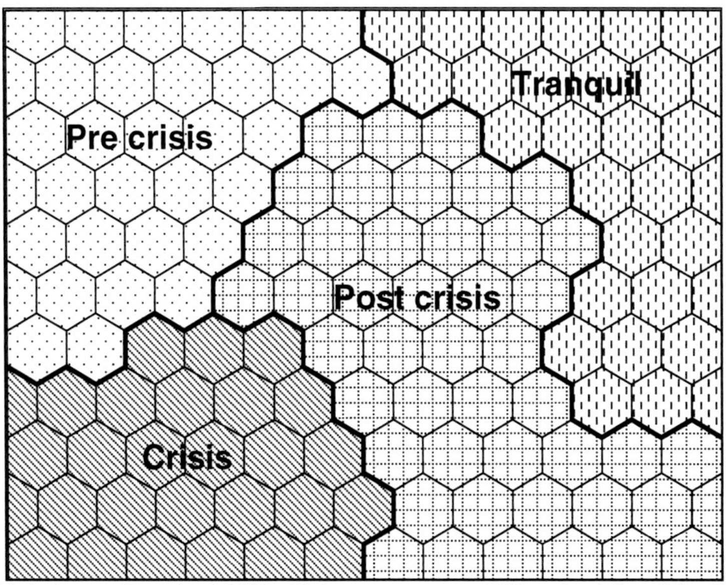

Notes: This is the two-dimensional SOM grid that represents a multidimensional financial stability space. The lines that separate the map into four parts are based on the distribution of the four underlying c. For distributions of individual indicators and class variables on the SOM, see Fig. 3.

Fig. 2. The two-dimensional SOM grid representing a multidimensional data space

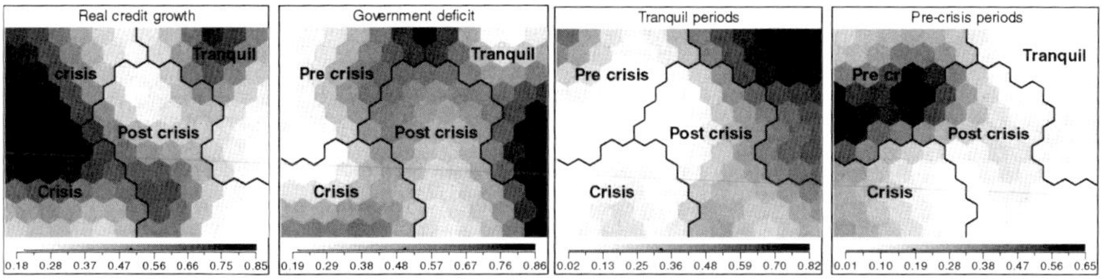

Fig. 3. Feature planes for two indicators and two class variables

To partition the map into the four stages of the financial stability cycle (pre-crisis, crisis, post-crisis and tranquil states), the second-level clustering of the map is performed using Ward's hierarchical clustering with respect to the class variables. To explore individual dimensions of the SOM model in Fig. 2, one can visualize individual layers. Fig. 3 illustrates examples of the feature planes (two indicators and two class variables). These can be used for assessing correlations of the vulnerability indicators with the classes for finding typical patterns during pre-crisis, crisis, post-crisis and tranquil stages of the financial stability cycle. For instance, the two example feature planes show that real credit growth is high during pre crisis and crisis periods and that government deficits are generally high during post-crisis and tranquil periods. The two latter feature planes show that class variables are, obviously, distributed in their respective clusters.

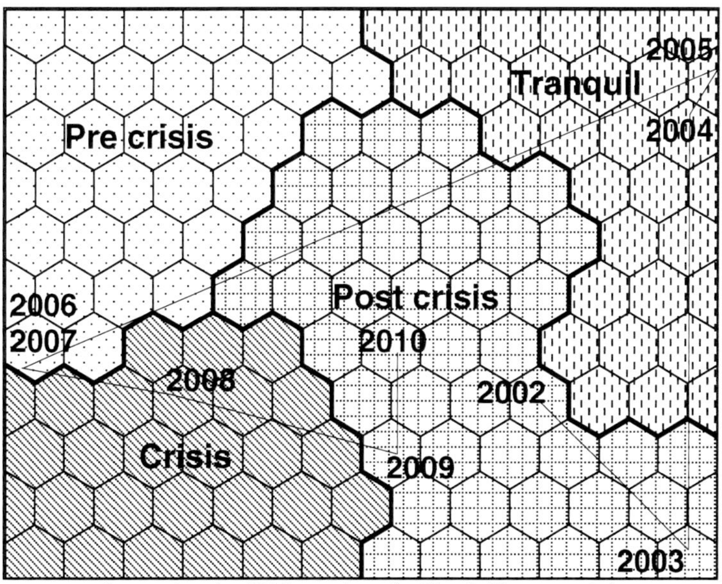

Notes: For notes on interpreting the SOM, see those for Fig. 2. The mapped data are macro-financial vulnerabilities representing the first quarter per year from 2002–2010 for Japan.

Fig. 4. Mapping the state of financial stability in Japan from 2002–2010 on the SOM

The two-dimensional SOM grid (Fig. 2), which represents a multidimensional financial stability space, can now be used for mapping economies across countries and over time. Fig. 4 demonstrates the evolution of macro-financial vulnerabilities in Japan from 2002–2010. While each data point is mapped to its best-matching unit, the time dimension is shown by connecting consecutive data. The figure shows that, during the analyzed period, Japan moves through all four stages of the financial stability cycle.

3.2 Discovering Chances of Financial Imbalances with the SONM

In this section, we use the Self-Organizing Network Map (SONM) for two purposes: for monitoring contagion possibilities of country-level financial stability using financial linkages and for assessing chances of rare events using the frequency distribution on the grid.

The two-dimensional SONM grid visualizes two different but related types of data. We assess the evolution of financial stability over time and across countries as well as map bilateral financial linkages between cross sections of countries. Fig. 5 demonstrates the method by showing a projection of a time series for the United States as well as a cross section for the latest point, quarter 3 in 2010. In the figure, a black trajectory shows the evolution of annual macro-financial vulnerabilities in the US. The last data point, 2010Q3, represents the mid-point of the superimposed network of financial linkages. Portfolio exposures are based upon external assets (equities and bonds) as reported in the Coordinated Portfolio Investment Survey by the IMF. Exposures of central banks are not included due to the different nature of their holdings. Node size of each country is scaled as to the sum of exposures to other countries. The thickness of the edges represents the size of external exposure to total exposures per country, where the color of the edge indicates the address of the exposure holder. The map not only demonstrates how the US development evolves through all four stages of the financial stability cycle, but also presents the size of possible contagion channels of instabilities to and from the US economy. Indeed, the SONM combines the state of financial stability, or probability of a crisis, with the systemic relevance of the economy.

Fig. 5 shows a number of interesting attributes of the cross section in 2010Q3. Based upon the location of the countries linked to the US, we find four considerably distinct clusters of financial stability states: two groups of stable developed economies in the lower right corner of the grid and two more vulnerable groups of less developed economies in the upper left corner of the grid. While the countries in the upper part of the map are generally emerging market economies, the countries in the lower part are mixed but mainly advanced economies. This suggests that in 2010Q3 risks were higher in emerging markets. The events in 2011 have, however, turned the attention to advanced economies. The size of financial linkages to high-risk economies (e.g. Argentina, Brazil, Mexico etc) enlightens about the past, present and future: not only may high levels of previous financial stress in the US have impacted on their current state, but they also have a high risk of current and future shock propagation from the US. This enables discovering those economies that are particularly vulnerable to financial instabilities in other regions. While crises are often transmitted through contagion channels like financial linkages, they may also be propagated through

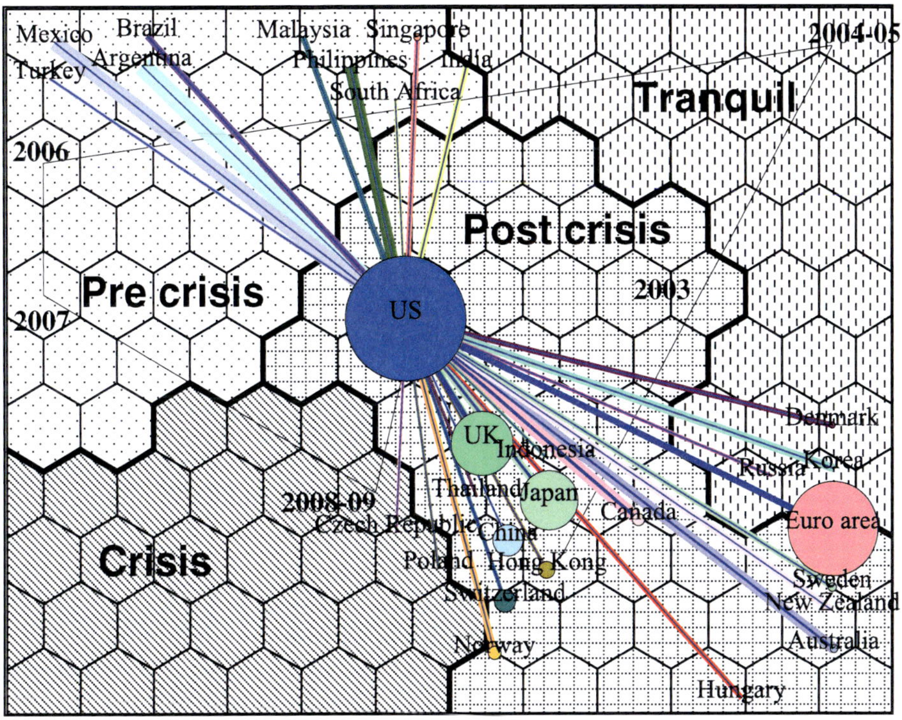

Notes: The figure displays the two-dimensional SONM. It superimposes financial linkages on top of a SOM grid as well as a trajectory for the United States from 2002 to 2010 (blue node in the centre).

Fig. 5. Discovering chances in a financial network with the Self-Organizing Network Map

similarities in macro-financial conditions, something extremely important when dealing with data of changing nature [13]. Independent of location on the map, an economy that is located in the neighborhood of countries in crisis could through contagion experience a similar wave of financial distress. Thus, an economy in the upper left part of Fig. 5, say Brazil, could propagate financial instabilities to countries with similar macro-financial vulnerabilities, i.e. Argentina, Mexico and Turkey.

The data topology of the SOM can also be related to another common task of Chance Discovery – finding events bridging multiple clusters of co-related frequent events. As the SOM attempts to maintain neighborhood relations in data on a grid, rare nodes (or clusters) neighboring (or linking) frequent ones can be regarded as candidate chances which have been rare in the past but may have a high impact in and on the future. By looking at the frequency distribution of multivariate data on the SOM grid, we can use the model for assessing chances of rare, but significant, events. The left grid in Fig. 6 shows the frequency distribution of the SONM. It shows, for instance, that the borders of the clusters, in particular the outer borders, have high frequencies, while the centers tend to have lower frequencies. As the SOM approximates the probability density function of data, the nodes are located in more dense parts of the data space. Hence, the U-matrix, or a non-linear projection technique aiming at preserving global distances like Sammon's mapping [20], is an ideal extension for finding rare, or extreme, locations in the data space through bilateral node-to-node distances. For monitoring predefined states in data, as is done here, this exercise is not as interesting as for tasks with entirely unknown relationships. In various forms of customer relationship management, for example, infrequent target markets (nodes) neighboring frequent ones on profitable parts of the grid could be ideal targets for direct marketing.

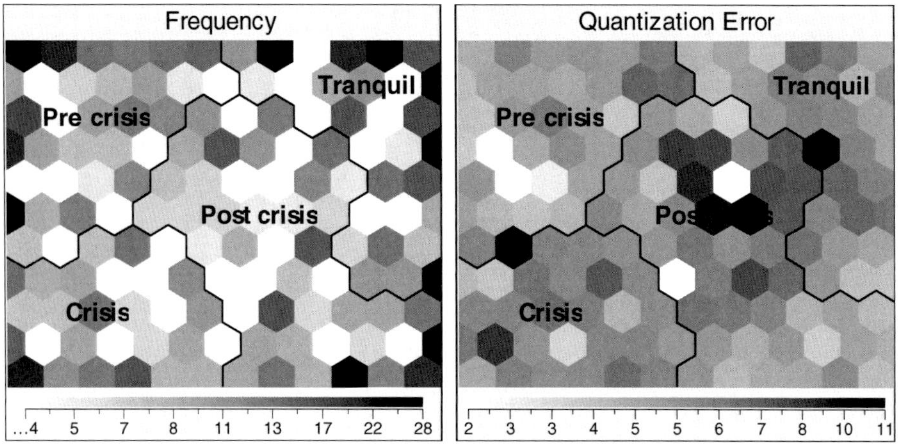

Fig. 6. Frequency and quantization error of data in each of the nodes

The right grid in Fig. 6 shows, for each node, the average distance between node m_i and all data x that belong to node m_i (i.e. best-matching unit, BMU). When comparing the grids representing frequency and quantization error, one can easily observe that they are to some extent negatively correlated. This may reflect the fact that rare events are somewhat heterogeneous in nature. It is also important to note that rare events (nodes) are highly affected by neighbors, which might distort the distance. The quantization error can be seen as quantization quality in general or as the correspondence between the prototypes (nodes) and the data in particular. However, a more meaningful estimate of event rarity is computing the distance of individual data points x_j to their BMUs. As the nodes m_i of the SOM, while being topologically ordered, tend to approximate the probability density function $p(x)$ of data x [9], the distance to a BMU represents the fit of a single data point to the multivariate data distribution. An outlier, and its degree of extremity, can thus be estimated by the distance to the SOM. When monitoring financial stability, one could assume that large distances represent financial imbalances in the macro-financial conditions. However, the imbalance, being it in a post-crisis, crisis or pre-crisis state, provides only information of impending instabilities rather than information on the exact timing of a crisis.

The SONM is a straightforward extension for assessing imbalances in cross-sectional financial networks, and could obviously as well be applied in a broad scope of other fields with links in the cross-section. As is said, the extension of a network analysis only facilitates discovery of static chances. For temporal data, there exist several techniques that enhance both the representation and the prediction of the SOM. The trajectory in Fig. 3 is a common means for individual analysis. In Sarlin and Eklund [21–22], this was extended to show membership degrees of each time-series point to each cluster. Probabilistic modeling of state transitions on the SOM has been shown to facilitate exploiting strengths and actual directions of temporal patterns in financial data [23]. SOM processing has also been adapted for time-series prediction (see e.g. [24-25]). These extensions for time-series data are, similarly as the SONM for cross-sectional analysis, suitable for temporal tasks in Chance Discovery, as they use the ordering of nodes as a basis and link various visualizations to that.

4 Conclusions

In this chapter, we have discussed the Self-Organizing Map (SOM) within the scope of Chance Discovery. The SOM paradigm supports several principal parts of Chance Discovery: visualization of temporal and cross sectional multivariate data, discovering rare clusters bridging frequent ones, outlier and event rarity analysis, and dealing with continuously evolving structures of real world data. We further enhance the standard SOM paradigm by combining it with network analysis. It enables a simultaneous view of the data topology of the SOM and a network topology of relationships between objects on the SOM. The usefulness for Chance Discovery of the Self-Organizing Network Map (SONM) is demonstrated on a dataset of macro-financial measures. While the macroprudential indicators show country-specific vulnerabilities by positions on the SONM grid, the financial linkages show the size of exposure and potential shock propagation to and from the rest of the world. The case study demonstrates a successful application to financial stability monitoring. Indeed, the SOM, and the SONM in particular, could as well be used for Chance Discovery in a broad range of other real-world applications, such as customer relationship management and earthquake-risk detection.

Acknowledgments. The author wants to thank Barbro Back, Tomas Eklund and Tuomas Peltonen, and seminar participants at the International Joint Conference on Artificial Intelligence (IJCAI'11) workshop on Chance Discovery on 16–22 July 2011 in Barcelona and the Data Mining and Knowledge Management Laboratory at Åbo Akademi University for useful comments and discussions.

References

1. Ohsawa, Y.: Chance Discovery for Making Decisions in Complex Real World. New Generation Computing 20(2), 143–163 (2002)
2. Tsang, E.P.K., Markose, S., Er, H.: Chance discovery in stock index option and future arbitrage. New Mathematics and Natural Computation 1(3), 435–447 (2005)

3. Ohsawa, Y.: Modelling the process of chance discovery. In: Ohsawa, Y., McBurney, P. (eds.) Chance Discovery, pp. 2–15. Springer, Heidelberg (2003)
4. Quinlan, J.R.: C4.5: Programs for Machine Learning. Morgan Kaufmann, California (1993)
5. Greenacre, M.J.: Correspondence Analysis in Practice. Chapman & Hall, London (2007)
6. Ohsawa, Y., Benson, N.E., Yachida, M.: KeyGraph: Automatic Indexing by Cooccurrence Graph based on Building Construction Metaphor. In: Proc. Advanced Digital Library Conference, pp. 12–18. IEEE Press, Los Alamitos (1998)
7. Abe, A., Hagita, N., Furutani, M., Furutani, Y., Matsuoka, R.: An interface for medical diagnosis support. In: Apolloni, B., Howlett, R.J., Jain, L. (eds.) KES 2007, Part II. LNCS (LNAI), vol. 4693, pp. 909–916. Springer, Heidelberg (2007)
8. Kohonen, T.: Self-organized formation of topologically correct feature maps. Biological Cybernetics 66, 59–69 (1982)
9. Kohonen, T.: Self-Organizing Maps. Springer, Berlin (2001)
10. Ultsch, A., Siemon, H.P.: Kohonen's self organizing feature maps for exploratory data analysis. In: Proceedings of the International Conference on Neural Networks, pp. 305–308. Kluwer, Dordrecht (1990)
11. Vesanto, J., Alhoniemi, E.: Clustering of the self-organizing map. IEEE Transactions on Neural Networks 11(3), 586–600 (2000)
12. Ward, J.: Hierarchical grouping to optimize an objective function. Journal of the American Statistical Association 58, 236–244 (1963)
13. Matsuo, Y.: Prediction, Forecasting and Chance Discovery. In: Ohsawa, Y., McBurney, P. (eds.) Chance Discovery, pp. 30–43. Springer, Heidelberg (2003)
14. Boulet, R., Jouve, B., Rossi, F., Villa, N.: Batch kernel SOM and related Laplacian methods for social network analysis. Neurocomputing 71(7-9), 1257–1273 (2008)
15. Goda, S., Ohsawa, Y.: Chance Discovery in Credit Risk Management - Time Order Method and Directed KeyGraph for Estimation of Chain Reaction Bankruptcy Structure. In: Satoh, K., Inokuchi, A., Nagao, K., Kawamura, T. (eds.) JSAI 2007. LNCS (LNAI), vol. 4914, pp. 247–254. Springer, Heidelberg (2008)
16. Sarlin, P., Peltonen, T.A.: Mapping the State of Financial Stability. ECB Working Paper, No. 1382 (September 2011)
17. Lo Duca, M., Peltonen, T.A.: Macro-Financial Vulnerabilities and Future Financial Stress — Assessing Systemic Risks and Predicting Systemic Events. ECB Working Paper, No. 1311 (2011)
18. Borio, C., Lowe, P.: Asset Prices, Financial and Monetary Stability: Exploring the Nexusd. BIS Working Papers, No. 114 (2002)
19. Borio, C., Lowe, P.: Securing Sustainable Price Stability: Should Credit Come Back from the Wilderness? BIS Working Papers, No. 157 (2004)
20. Sammon Jr., J.W.: A Non-Linear Mapping for Data Structure Analysis. IEEE Transactions on Computers 18(5), 401–409 (1969)
21. Sarlin, P., Eklund, T.: Fuzzy Clustering of the Self-Organizing Map: Some Applications on Financial Time Series. In: Laaksonen, J., Honkela, T. (eds.) WSOM 2011. LNCS, vol. 6731, pp. 40–50. Springer, Heidelberg (2011)
22. Sarlin, P., Eklund, T.: Financial Performance Analysis of European Banks using a Fuzzified Self-Organizing Map. In: König, A., Dengel, A., Hinkelmann, K., Kise, K., Howlett, R.J., Jain, L.C. (eds.) KES 2011, Part II. LNCS, vol. 6882, pp. 186–195. Springer, Heidelberg (2011)

23. Sarlin, P., Yao, Z., Eklund, T.: Probabilistic Modeling of State Transitions on the Self-Organizing Map: Some Temporal Financial Applications. In: Proc. of the 45th Hawaii International Conference on System Sciences, HICSS 2012 (forthcoming, 2012)
24. Chappell, G., Taylor, J.: The temporal Kohonen map. Neural Networks 6, 441–445 (1993)
25. Strickert, M., Hammer, B.: Merge SOM for temporal data. Neurocomputing 64, 39–72 (2005)

Map Interface for a Text Data Set by Recursive Clustering

Wataru Sunayama, Shuhei Hamaoka, and Kiyoshi Okuda

Graduate School of Information Sciences, Hiroshima City University
3-4-1 Ozuka-Higashi, Asa-Minami-Ku, Hiroshima, 731-3194, Japan
{sunayama}@sys.info.hiroshima-cu.ac.jp

Abstract. Recently, there have been many opportunities to acquire text information as the quantity of electronic information increases. Data classification or clustering methods are widely adapted in order to acquire various information effectively from an enormous active text data set. However, ordinal clustering methods connect texts and many texts are concentrated into a single cluster so that we cannot see various information.

In this study, we propose a recursive clustering method to avoid such bias by integrating a set of texts, included in a cluster, into a single text. An interface that we can comprehend a result of clustering intuitively and can explore information is required to grasp an overview of data and to be led to a new idea. According to the experimental results, the proposed method could construct clusters that are not biased. Test subjects could find information widely by using a map visualizing clustering results.

Keywords: map interface, information collection, recursive clustering, information visualization.

1 Introduction

Recently, there have been many opportunities to acquire text information as the quantity of electronic information increases. Data mining, text mining and chance discovery[1] techniques that can assist us to notice valuable insights and to connect to new ideas are required. In order to find new clues sealed in enormous data as chance discovery, we must grasp the whole image of data and classify them to overview. Data classification or clustering methods are widely adapted in order to acquire various information effectively from an enormous active text data set. However, ordinal clustering methods connect texts and many texts are concentrated into a single cluster so that we cannot see various information.

Especially, when there are no categories to be classified for an active text data set, clustering methods must focus on its dispersion of clustered data rather than its accuracy. That is, we have a goal to find new ideas by seeing whole data and those meanings, though many ordinal clustering methods considers how classifies data correctly the most.

Y. Ohsawa & A. Abe (Eds.): Advances in Chance Discovery, SCI 423, pp. 63–77.
springerlink.com

In this study, we propose a recursive clustering method to avoid such bias by integrating a set of texts, included in a cluster, into a single text. An interface that we can comprehend a result of clustering intuitively and can explore information is required to grasp an overview of data and to be led to a new idea. That is, the following supports are proposed.

1. Improve visibility by visualizing hierarchical data structure constructed by a novel recursive clustering.
2. Display data with a map interface that we can comprehend data intuitively.
3. Display data with a map interface that we can operate to explore data intuitively.

2 Map Interface by Recursive Clustering

Fig.1 shows the proposed framework for constructing a map by recursive clustering. The system requires texts as input, the system calculates relationship values between texts at first. Links are created between texts, and clusters are created by the links. If more than two clusters exist, texts in a cluster are integrated into a text and the system creates clusters recursively. Such an iterative process creates a hierarchical clustered structure. Finally, a map is created expressing the constructed hierarchical structure and output.

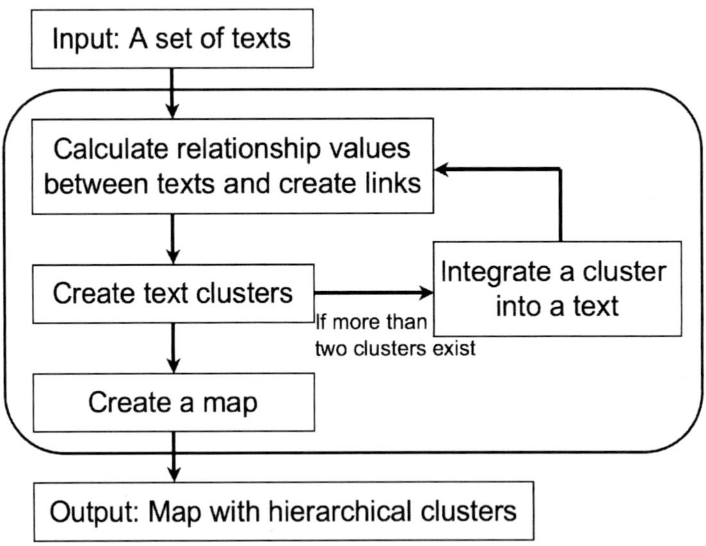

Fig. 1. Framework for constructing a map by recursive clustering

2.1 Input: A Set of Texts

Texts including at least one noun can be treated as input. For example, a set of texts obtained as a search result can be an input [1].

2.2 Recursive Clustering

The recursive clustering process consists of three processes, calculation of relationship values between texts for creating links, clustering, and integration of texts in each cluster. Each process is described in the following subsections.

Calculation of Relationship Values and Link Creation. Relationship values between texts are calculated by the cosine similarity given as Eq.(1), where $Num(a)$ denotes the number of words/nouns included in a text a.

$$Relation(a, b) = \frac{Num(a \cup b)}{\sqrt{Num(a) \times Num(b)}} \tag{1}$$

Links are created between texts whose relationship values are more than the threshold K. Though the number of recursive clustering depends of the threshold K, K is set as the maximum value while the number of links do not exceed the half number of texts. For example, when the number of texts is 100, K is set as the maximum value where the number of links do not exceed 50.

Clustering Method. Newman method [12] is employed as a clustering method. The algorithm is as follows.

1) Define each text as a cluster.
2) Combine two clusters whose difference of evaluation value Q, that is ΔQ, become the maximum.
3) Repeat the step 2) until the value Q become the maximum.

Evaluation value Q is defined as Eq.(2), and its differential value ΔQ is expressed as Eq.(3), where e_{lm} denotes the number of links connecting two texts that belong to clusters l and m each.

$$Q = \sum_{l} (e_{ll} - {a_l}^2) \tag{2}$$

$$\Delta Q = e_{lm} + e_{ml} - 2a_l a_m = 2(e_{lm} - a_l a_m) \tag{3}$$

$$a_l = \sum_{m} a_{lm} \tag{4}$$

This clustering method creates clusters so that the number of links connecting texts belong to the same cluster will increase and the number of links connecting texts that belong to different clusters will decrease.

[1] The number of texts should be less than 1,000 currently because the algorithm of the process is not well considered. In the future, considering visibility of an output map, 10,000 texts can be available in this environment.

Integration of Texts in a Cluster. This process integrates texts in each cluster into a text. That is, texts in a cluster are all combined into a text by connecting those texts. It is needless to consider the connecting order because only frequency information is used from each text to calculate relationship.

Recursive Clustering Method. In the recursive clustering process, the amount number of clusters gradually decrease as clusters are integrated. Finally, when the all input texts were integrated into a single text, or when the threshold K became 0.01 [2], the recursive process would be terminated.

Ordinal many clustering methods tends to create a big cluster that contains many elements especially when a set of texts that are related one another is given. In such cases, the visibility of texts are sacrificed in front of the large mass of data. In the proposed method, even if a large cluster was created once, the texts included in the cluster would be combined into a single text and the weight of the cluster would not become large. On the other hand, this method has a risk to classify texts incorrectly.

2.3 Map Creation

In this section, a drawing method to create a map expressing hierarchical structure is described.

The drawing methods is an extended one of a map animation interface [11]. Primitives in the map are shown in Table 1. The primitives except for "Area" have been already used, so the "Area" is newly introduced because "Island" that denotes cluster should be piled up in a hierarchical clustering structure.

Clustering results are expressed as "Islands," "Rails," "Roads," and "Towns" corresponding to clusters, strong links, weak links, and texts respectively when recursive process is not incorporated. Recursive clustering process is expressed as dendrogram, so reversed one like Fig.2 is used for creating a map. That is, each cluster except for the bottom and the second bottom of the Fig.2 called an area. Areas are piled up on islands and islands are on the sea.

The names of islands and areas are extracted from the texts included in those islands or clusters and attached. The names are simply chosen by the document frequency, the number of the documents appearing a specific noun, and the most appeared nouns are extracted as the candidates of labels in the order of frequency. Then, the labels are given to islands at first, and given from the bottom to the top in Fig.2. In this process, if a noun is used as a label once, that noun is not used twice in the direct sequence of the dendrogram, connected as an ancestor and a child.

Fig.3 show an overview of the interface. The background of the map expresses the sea and the map contains three islands that includes piled areas with white lined borders.

[2] K started from 1.00 and decreases by 0.01.

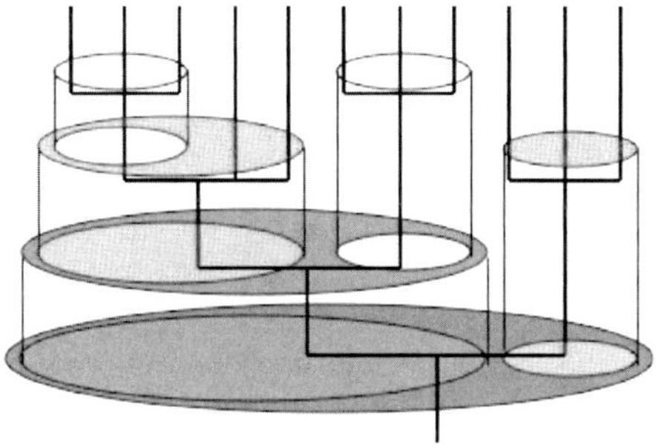

Fig. 2. Islands construction by a reversed dendrogram

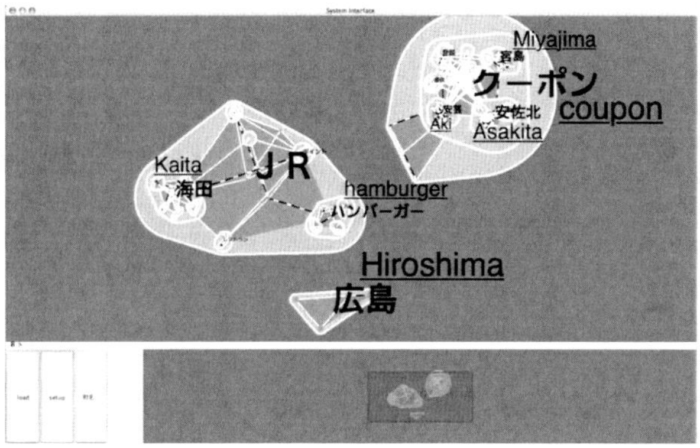

Fig. 3. Interface overview

2.4 Example Use of Interface

Fig.3 consists of the main map and the sub map that are used for global and local views.

Users basically use the main map using operating commands "Zoom in," "Zoom out," "Translation," and "Selection" as in Table 2. That is, users can use the map like a on-line map service, and "Selection" is realized by double clicks to a town with a mouse to see concrete text information.

For example, suppose a user who wants to have lunch near his university. That user gave texts about shops for lunch as input to the system, and a map is displayed like top left of Fig.4. The user thought that he would like to have lunch near Japanese Railway station, so he zoomed in the island "JR" by using a mouse wheel with translation operation as in the top center of Fig.4. As he saw the "JR" island, he found "Hamburger" area and zoomed in again as in the right of Fig.4. After that, he zoomed in the "Shop" area and double clicked the town "Midorii," the name of place, and accessed an concrete information like the bottom left of Fig.4.

Table 1. Primitives for the proposed map interface

Name	Image	Explanation
Islands	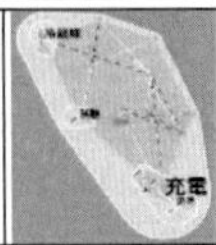	Areas surrounded by the sea, the biggest clusters. An island is a set of areas and texts.
Areas	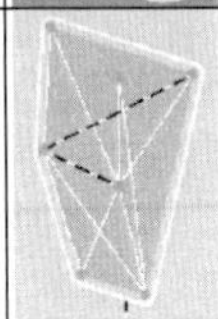	Areas included in an island, clusters. An area is a set of areas and texts.
Towns		Round nodes. A town is a text.
Roads		Roads are lines connecting towns and areas. The relationship values are more than the threshold.
Rails		Rails are zebra lines connecting towns and areas. The relationship values are more than 0.97.

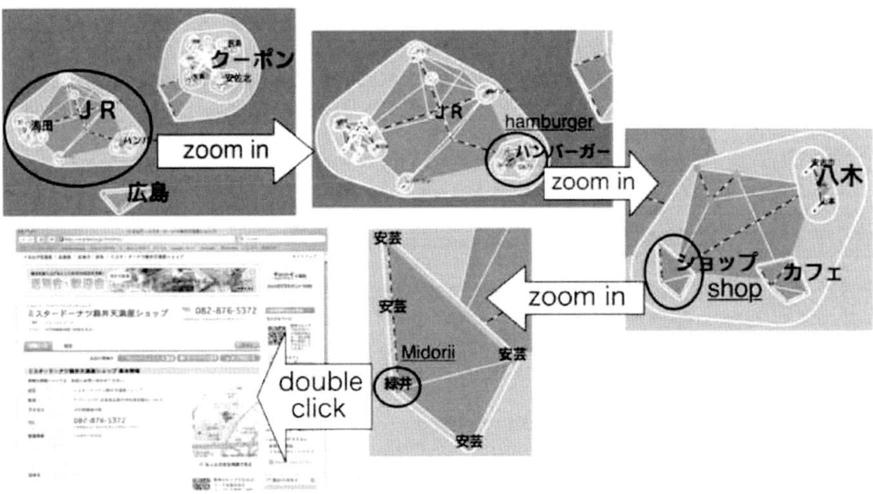

Fig. 4. Example operations

3 Cluster Density Evaluation

In this chapter, we describe the results of examination whether texts in clusters created by the proposed method are scattered or not.

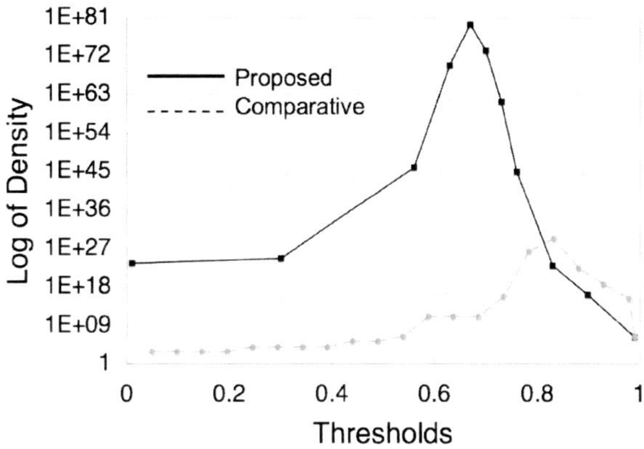

Fig. 5. Densities for the theme "Car"

Table 2. User operations in map interface

Operation commands	User mouse operation
Zoom in	Scroll up the mouse wheel
Zoom out	Scroll down the mouse wheel
Translation	Dragging
Selection	Double clicks

Table 3. Texts used for density evaluation

Themes	Kind of information	Texts
1. Drink	Restaurant information	961
2. Car	Japanese car information	1021
3. World food	Food information	256
4. Game soft	Reviews for games	502
5. Horror	Horrible stories	495
6. Company	Panasonic goods information	626

3.1 Evaluation Method

The examination was executed by comparison between two clustering methods, the proposed method and the minimal distance method. The minimal distance method combines the most related clusters recursively by treating the relationship value as the most valuable links between two clusters. The data used were shown in Table 3 downloaded from Web sites.

Density value *Density* is defined as Eq.(5) for this examination, where $node(i)$ denotes the number of texts in the cluster i.

$$Density = \prod_i node(i) \tag{5}$$

That is, the density value *Density* become large when the number of data and clusters are the same and the data are scattered equally. In this examination, as the number of data were set as the same, so the method whose value was large could create clusters with no polarization.

3.2 Evaluation Results

From Fig.5 to Fig.6 show the results in semi-log graphs. The vertical axis means log of density *Density* and the horizontal axis means the threshold to create links.

According to these figures and results for the other texts, the maximum values in those graphs in the proposed method became very larger than those of the comparative method. Therefore, it is confirmed that the proposed clustering method could disperse data more widely. The values of the left ends were also larger in the proposed method. Since left ends mean the most biggest items, corresponding to islands in the map, so it is expected that we can find various information.

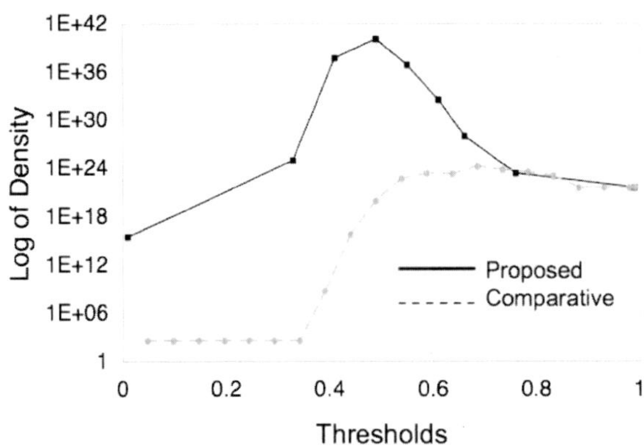

Fig. 6. Densities for the theme "Game soft"

Table 4. Texts for precision evaluations

Sites	Categories	Number of texts
News-1	Culture, Economy, Entertainment, Sports, etc.	1000
News-2	Movies, Novels, Fashions, Restaurants, etc.	1000
Reviews	Cell phones, Cars, Watches, Games, etc.	1000

4 Precision Evaluations of Information Clustering

In this chapter, we describe the evaluation results of precisions and F-measures for the proposed recursive method by comparing an existing clustering method.

Since the proposed method integrates texts in a cluster on the process of recursive clustering, the effects of low frequency words become relatively strong. Though texts were dispersed in many clusters as in the evaluation of the last chapter, precisions of clustering seem to be decreased compared to the original clustering method. While the main purpose of the proposed method is to disparse information in order to collect various information, precisions of clustering must be kept in some certain level. Therefore, clustering precisions of the proposed and Newman's method[12] were evaluated by the following procedures with three existing text database.

A thousand of texts in ten categories, a hundred each, for each Web site were prepared as in Table 4. These categories were chosen as many of them were not overlapped. The proposed and Newman's clustering methods were applied to the prepared database to produce precisions and recalls. When we calculated precisions and recalls, one cluster was selected for evaluating one category as the

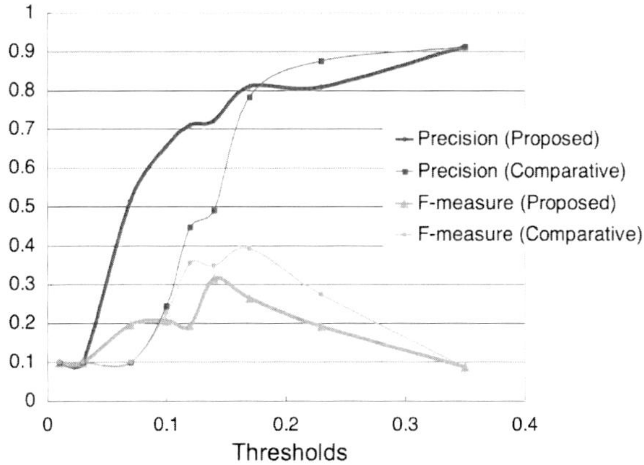

Fig. 7. Precisions and F-measures for the theme "News-1"

F-measure, harmonic mean of precision and recall, became the highest [3]. Each system was evaluated by the average produced by the values for ten categories.

From Fig. 7 show precisions and F-measures of clustering results. Though the maximum values of F-measures were larger in Newman's method, those of the proposed method did not become worse remarkably. In the lower parts of horizontal axis, thresholds to create links between texts, precisions of the proposed methods were larger than those of Newman's. This implies that the proposed method created plural clusters in the situation that Newman's method created a large cluster with the lower thresholds and many links.

According to these results, it is confirmed that the proposed recursive clustering method has a possibility to support our information collection.

5 Experiments: Information Collection by Using Map Interface

In this chapter, the experiments that testify how users can collect information widely with the proposed interface are described.

5.1 Experimental Settings

The texts in Table 3 were used for this experiment. The test subjects were 18 university and graduate school students who majored information sciences. Each test subject collected information according to the given themes such as "You have decided to examine goods whether the company is promising or not. Please pick up goods that express the company well." in the theme "Company." That is, themes for information collection were given to the subjects.

The comparative clustering method was the minimal distance method used in the last examination. The map interface was common in two methods. Each subject collected information six times; three times for each. The order of answer and assignment were arranged properly so as not to have biases.

The evaluation should be executed how wide the subjects could collect information. Therefore, *Dispersion* value was defined as the averaged number of meeting points on the dendrogram created by the comparative method. Meeting point is defined as the place where the two texts were combined on the dendrogram. The dispersion value for one meeting point is counted from leaves as zero and increase one by each iterative clustering. That is, the dispersion values become large if collected data were scattered in the dendrogram.

5.2 Experimental Results

Fig.8 shows the dispersion values for the subjects who collected information positively, and Fig.9 shows the dispersion values for the subjects who collected information negatively. "Positive" means the subjects who collected information

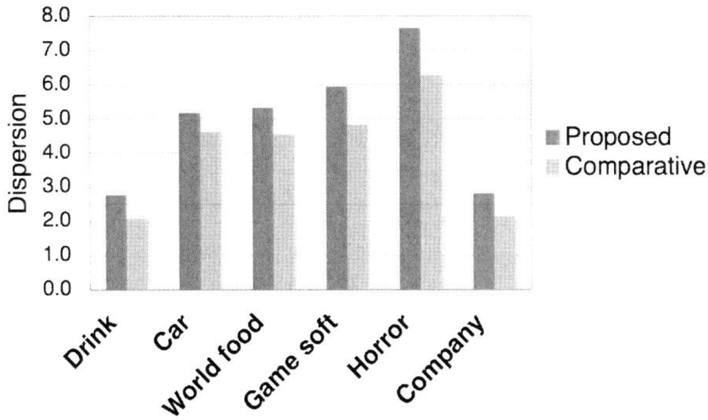

Fig. 8. Dispersions for the subjects positively collected

more than the average in each theme, and "Negative" means the subjects who collected less than the average.

According to the Fig.8, *Dispersions* of the subjects who used the proposed system could collect more information than those who used the comparative system. This is interpreted that the proposed clustering method scatters texts into many clusters so that the subjects could see various islands or areas. On the contrary, Fig.9, the results for the negative subjects, says that the *Dispersions* were almost the same except for the theme "Horror." This is because, information was tend to be retrieved from distinctive areas naturally if the quantity of collected information was small. As a result, when a user would like to collect many information, the proposed system is available for making an effective collection.

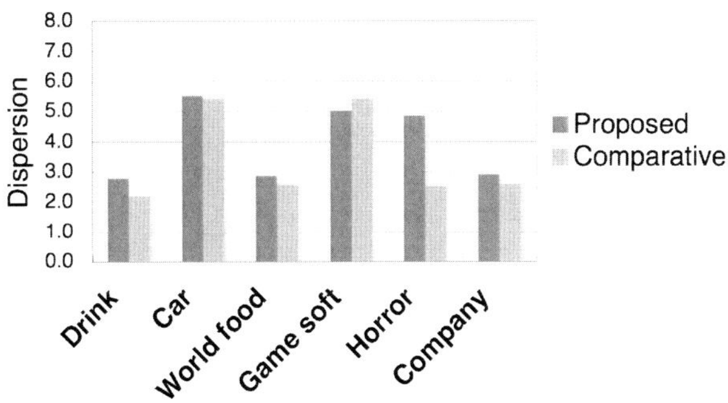

Fig. 9. Dispersions for the subjects negatively collected

Table 5 shows the numbers of collected information. The subjects who used the proposed method could collect many information than those with the

[3] It was adjusted that one cluster was not used twice for evaluating two categories.

comparative interface. Therefore, the proposed system could utilized to acquire many and various information.

Table 6 shows concrete information that collected by two subjects who used the proposed system and the comparative one each in the theme "Company." Both the subjects collected eight information. Though the subject who used the proposed system could collect different types of goods information, two VIERA and three LUMIX are contained in the information collected by the subject who used the comparative system. Therefore, it seems that the comparative method could not divide information widely.

It is interpreted that the test subjects could see various data as they could overview the data better than to use a comparative classification technique. That is, in order to grasp the whole image of data, tendencies and structures are more important than the accuracy of classification for accomplishing the initial purpose of this study.

While we have just confirmed that the proposed environment could assist users to overview data, chance discovery needs data comparison methods from various viewpoint. Therefore, studies to compare maps from different viewpoints and to encourage users to notice[11,13] have been proposed.

Table 5. Numbers of collected information

Theme	Proposed	Comparative
Drink	6.0	6.0
Car	6.4	4.2
World food	7.0	5.4
Game	7.0	5.8
Horror	4.3	3.1
Company	5.3	4.9
Average	6.0	4.9

Table 6. Titles of collected informations in the theme "Company"

Proposed method (Dispersion: 2.6)	Comparative method (Dispersion: 1.8)
C3 Liquid crystal television VIERA	Liquid crystal television VIERA
Digital camera DMC-ZX3	X3 Liquid crystal television VIERA
Refrigerator	DMC-TZ LUMIX
Jar pot	DMC-FZ100 LUMIX
Jar Rice cooker	LUMIX Series G, Movie a single lens
Washing machine	Mini compo
Line up for facsimile	Microwaves
Massage chair / Foot massager	Fitness JOBA
Buttery	Air purifier

For this purpose, i.e., in order to compare cluster-based structures of different datasets, it is not meaningful to emphasize clustering performance in the sense of accuracy of items' belonging clusters. Instead, qualitative labels should be well assigned to clusters and the links among clusters are desired to be visualized to encourage intuitive understanding of users. According to the experimental results above, the method we proposed in this chapter satisfies this desire i.e., not meaninglessly accurate but meaningfully labeled and visualized.

In addition to this, the proposed method can be applied to compare structures, that is, we can discover differences and find remarkable points in structures from different viewpoints.

For instance, when a person who wants to compare article sets from each news Web site, he/she can analyze differences of tendencies by hierarchical structures of article sets. He/she also can investigate how each site treats the articles of a specific cluster, such as baseball news articles, by seeing the position, the size or lower structures of the cluster, If he/she was wanted to develop a new news site, the methods could be available to seek policy directions by comparing the treatment of each category articles.

As for the "Company"' appeared in the above experiment, job seeking students cannot see the all company to analyze in a real time. They can find the features and originality of each company by utilizing this comparing method. In the contrary case, some employees might be able to seek new breaking fields by comparing tendencies of many companies.

Table 7. Questionnaire results for usability and visibility

	Average (Max : 5 points)	
	Proposed	Comparative
Usability	4.4	4.7
Visibility	3.3	2.7

Since a person who discovers from data is not a computer but a human, methodologies of data arrangement and awareness support are nominated as ones of the most important subjects of chance discovery. Therefore, such an information restructuring method and a visualization method become imperative elements that form chance discovery. We need to continue to analyze and improve the information exploration process described in this chapter in order to design effective interaction between visualization interface and human.

Table 7 shows questionnaire results about visibility and usability. Usability was good in both methods because the map interface was intuitive and could be used naturally. However, visibility was not so good in both because the town names and area names were hard to be recognized occasionally. Therefore, the display algorithm should be improved in the future.

6 Related Works

6.1 Text Classification and Text Clustering

Studies for text classification have been executed since long ago, such as methods with Naive Bayes [2], with EM algorithms [3], with support vector machine [4] and with kernels of feature spaces [5] had been proposed. There are methods to classify texts not only by the original texts but by the labels or tags attached to the texts [6]. Though these methods are available to construct a database automatically, or are used for information retrievals with search engines, more structured clustering results are required to see the whole data and grasp the tendency of a text data set.

6.2 Visualization of Hierarchical Clustering Results

Visualization methods that express hierarchical data structure have been utilized for grasping whole data information and for acquiring information manually. There are methods to visualize a text data set with a hierarchical network, such as a method based on hyperbolic tree [7], data visualization on 3D sphere [8] or on a tree [9], and visualization as a map [10]. Though these methods focused on visualization of classification structure, the clustering methods being employed are major ones like K-means, so those methods do not adopt ideas that changes classification structure. That is, when a system creates clusters for a medium size of a text data set, a set has from 100 to 1000 that we might see all, related to a concrete topic, such text data in a set are not well divided because there are no basic categories to be classified. In such cases, many data are easily categorized in a single cluster and the structure itself become no longer useless.

Therefore, a method that combines a novel clustering method not oriented in its accuracy but in visibility and a map based intuitive interface is proposed in this study.

7 Conclusion

In this study, a recursive clustering method to improve visibility in order to collect various information is proposed. To visualize sophisticated structure is an essential point to grasp the whole data structure and find information to think of new ideas. Clues for new ideas are scattered in enormous information flood, so glittering data for someone are generally hidden. The precious data that are hard to be found do not appear by general visualization methods, so some points such as accuracy that is usually considered as important should be sacrificed in order to find new chances that are not found by anyone.

In future works, the system is expanded to visualize data that should be pointed out by some original measurement scores. Multiple data sets should be treated as inputs in order to display plural clustering results for comparing each structure. Intuitive data recognition and awareness must be realized as practical techniques of the chance discovery.

References

1. Ohsawa, Y., Nara, Y.: Decision Process Modeling across Internet and Real World by Double Helical Model of Chance-Discovery. New Generation Computing 21(2), 109–122 (2003)
2. McCallum, A., Nigam, K.: A Comparison of Event Models for Naive Bayes Text Classification. In: Proc. of AAAI-1998 Workshop on Learning for Text Categorization, pp. 41–48 (1998)
3. Nigam, K., McCallum, A.K., Thrun, S., Mitchell, T.: Text Classification from Labeled and Unlabeled Documents using EM. Machine Learning 39, 103–134 (2000)
4. Tong, S., Koller, D.: Support Vector Machine Active Learning with Applications to Text Classification. Journal of Machine Learning Research 1, 45–66 (2001)
5. Lodhi, H., Saunders, C., Shawe-Taylor, J., Cristianini, N., Watkins, C.: Text Classification using String Kernels. Journal of Machine Learning Research 2, 419–444 (2002)
6. Begelman, G., Keller, P., Smadja, F.: Automated Tag Clustering: Improving search and exploration in the tag space. In: Proc. of Collaborative Web Tagging Workshop at WWW (2006)
7. Kobayashi, T., Misue, K., Shizuki, B., Tanaka, J.: Information Gathering Support Interface by the Overview Presentation of Web Search Results. In: Proc. of Asia-Pacific Symposium on Information Visualisation, vol. 60, pp. 103–108 (2006)
8. Sprenger, T.C., Brunella, R., Gross, M.H.: H-BLOB: A Hierarchical Visual Clustering Method Using Implicit Surfaces. In: Proc. of the 11th IEEE Conference on Visualization (2000)
9. Akhavi, M.S., Rahmati, M., Amini, N.N.: 3D Visualization of Hierarchical Clustered Web Search Results. In: Proc. of International Conference on Computer Graphics, Imaging and Visualisation, pp. 441–446 (2007)
10. Muhr, M., Sabol, V., Granitzer, M.: Scalable Recursive Top-Down Hierarchical Clustering Approach with implicit Model Selection for Textual Data Sets. In: IEEE Workshops on Database and Expert Systems Applications, pp. 15–19 (2010)
11. Nishikido, T., Sunayama, W., Nishihara, Y.: Valuable Change Detection in Keyword Map Animation. In: Gao, Y., Japkowicz, N. (eds.) AI 2009. LNCS, vol. 5549, pp. 233–236. Springer, Heidelberg (2009)
12. Newman, M.E.: Fast algorithm for detecting community structure in networks. Phys. Rev. E 69 066133, 1–5 (2004)
13. Nohuddin, P.N.E., Christley, R., Coenen, F., Setzkorn, C., Sunayama, W.: Trend Mining and Visualisation in Social Networks. In: Proc. of the Thirty-first SGAI International Conference on Innovative Techniques and Application of Artificial Intelligence(AI 2011), pp. 269–282. Springer (2011)

Multimodal Discussion Analysis Based on Temporal Sequence

Katsumi Nitta

Interdisciplinary Graduate School of Technology and Science,
Tokyo Institute of Technology

Abstract. This research proposes a novel method for analysis of discussion record. One of the important features of our approach is to use both a logical analysis method and a word occurrence analysis method. A subject of discussion is analyzed and important issue factors are listed before the discussion starts. The logical analysis method describes the structure of the discussion referring to the issue factors. The word occurrence analysis method recognizes key topics and key utterance by observing utterances and nonverbal information such as action, facial expressions and so on.

1 Introduction

Discussion plays an important role in the resolution of disputes such as negotiation, moderation and arbitration. Discussion is modeled as exchanges of *arguments* on a specific topic. An argument is a pair of conclusions and their grounds which supports each conclusion. During discussion, an argument may be attacked by a counterargument. By exchanging arguments and counter arguments, discussion becomes more detailed and more complex. When a discussion includes a lot of topics (issue points) and they are related to each other, it sometimes becomes hard for participants to capture the whole structure of the discussion to understand which issue points are used to reach a consensus, and which arguments defeated other counter arguments. In such cases, a discussion support system which visualizes the structure of arguments and shows various features to evaluate the discussion skills will be helpful.

To support the analysis of discussion, a lot of research has been conducted so far. For example, some research done has represented the logical relationship among arguments in the form of a diagram, and has analyzed structure of a discussion [Reed 04]. Another research represents features of a discussion as a set of propositions, and estimates its conclusion by searching for similar cases from the past [Aleven 97]. Other research has focused on the statistics of the utterances that occurred during the discussion, and tries to find time points where topic change occurred [Ohsawa 06].

The researches done in the past have showed a useful effect for analyzing some aspects of discussion. However, they have been insufficient in anlyzing not only the structure of discussion, but also discussion moderation skills. For example,

Y. Ohsawa & A. Abe (Eds.): Advances in Chance Discovery, SCI 423, pp. 79–89.
springerlink.com

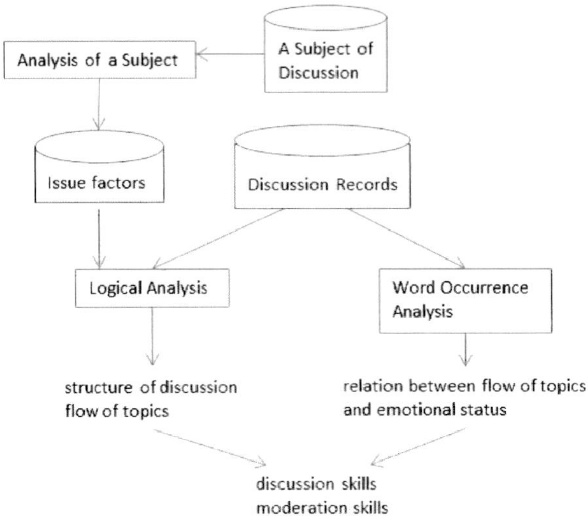

Fig. 1. Discussion Analysis Using Two Methods

when a discussion subject is decided, several topics to be discussed are estimated before the discussion starts. The subject includes several topics, and each topic is composed of several issue points. If the discussion is well moderated, these topics are discussed effectively. However, if the discussion skill is not of a high level, some important topics may be skipped or some topics may be discussed repeatedly. The emotional status of participants is also important for analyzing discussion moderation skills. If some participants cross their arms while speaking, he may be showing that he is irritated. In such a case, the chair person should change the topics or should take a coffee break considering the atmosphere of the discussion. Discussion moderation skills can be analyzed by extracting the logical structure and the emotional status from the discussion records and by comparing the discussion records with other discussion records whose subjects are the same. However, traditional analysis tools are not adequate to analyze discussion skills.

The objective of this research is to propose a novel method which supports the analysis of discussion record. One of the important features of our approach is to use both a logical analysis method and a word occurrence analysis method [?] [Maeno 06] [Nitta 09]. Fig. 1 shows the relation between the two methods. A subject of discussion is analyzed and important issue factors are listed before the discussion starts. The logical analysis method describes the structure of the discussion referring to the issue factors. The word occurrence analysis method recognizes key topics and key utterance by observing utterances and nonverbal information such as action, facial expressions and so on.

In Section Two, the logical analysis method is introduced. In Section Three, the temporal word clustering method and its extension to nonverbal information are introduced.

2 Logical Analysis of Discussion Records

A discussion is held based on the exchange of messages made by natural language. Although the same content is discussed, wording of the utterances and expressions vary depending on each individual speaker. For this reason, to compare several discussion records, we need to decide common factors to describe arguments which appear in the discussion.

We consider here that an argument is described using a proposition that indicates a fact or claim, and we call this proposition an *issue factor* (in this paper, we call it a *factor*). There exist a relationship where the establishment of one's factor supports (serves as the basis for) the establishment of the other one's factor, and a relationship where the establishment of one's factor attacks (conflicts with) the establishment of the other one's factor.

The following shows an example of the factors. In this case, f1 is the base of f3 (f3 holds because f1 holds), f2 is the base of f4 (f4 holds because f2 holds), while f3 conflicts with f4.

f1: The product sold was out of order.
f2: No malfunction was found in shipping.
f3: The seller is at fault.
f4: The seller is not at fault.

These relationships among factors can be expressed in an *issue graph* that is shown in Fig. 2. In this graph, each node corresponds to a factor, and a solid arrow shows a support relation between two factors and a dotted arrow shows an attack relation between two factors. This graph can be drawn when the main subject of a discussion has been determined before the discussion starts.

When a discussion starts, issues move onto this issue graph. Even though in discussions with the same subject, the transition of issues significantly varies depending on each individual speaker. By observing this transition, discussion moderation skills are evaluated.

To extract issue factors from utterances, morphological analysis should be conducted on the utterance messages and issues are estimated by utilizing combinations of words that occurred. We have proposed a machine learning method that discerns groups of words used for extracting issue factors from utterance records according to issues. Where multiple issue factors are extracted, an argument (a pair of conclusions and reasons) could be built depending on the combination of multiple issue factors. For example, when two issue factors, f1 and f3, are extracted and where there exists a relationship in which f1 supports f3 on the issue graph, this situation can be considered to represent the argumentation that "it is f3 because of f1."

As we described, even if the structure is the same, the process of discussion varies. For example, Fig.3 shows two flows of the topic of the discussion of the citizen judge system. The left figure is a topic flow by the lawyers, and the right figure is that of students. Lawyers topic flow is very effective because the same factor is not raised twice. On the contrary, right figure shows the same topic is brought up several times, which means this discussion is ineffective [Sato 11].

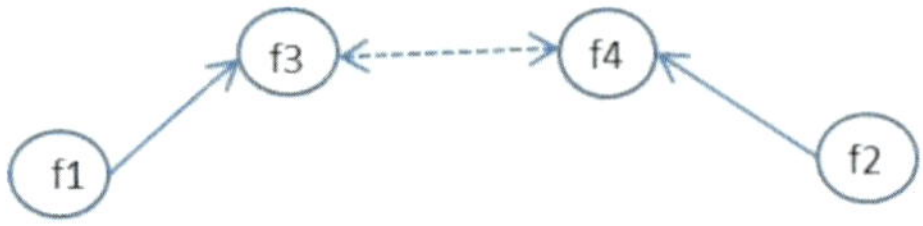

Fig. 2. Issue Graph

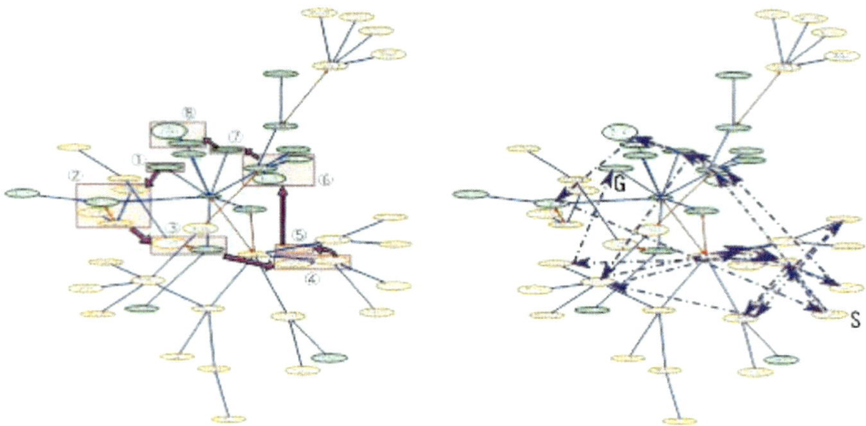

Fig. 3. Issue Graph and Topic Flow

3 Word Clustering Analysis

3.1 Basic Word Clustering Method

Maeno and Ohsawa regarded a discussion records as a set of utterance S1,S2,$\cdots$ Sm, and each utterance as a set of words {w1,w2,$\cdots$,wn} which occurred in the utterance [Maeno 09] [Maeno 06]. Distance between any two words (wi and wj) is defined using the Jaccard coefficient.

Given the intended number of clusters, all words are clustered using K-medoids method (Fig. 4). In Fig. 4, all words are clustered into three clusters (C1, C2 and C3). Each cluster is represented as a set of nodes and links. Each node corresponds to a word, and each link shows that Jaccard coefficient between these two words is a high score. We think that a cluster corresponds to a topic, so during discussion the focal cluster moves according to the change of topics.

After the clustering phase, in each utterance, a dummy node is inserted. A dummy node is different from other nodes because it doesn't represent a word, but it corresponds to an utterance. For each utterance, a ranking function is calculated. A ranking function measures the numbers of clusters which occurred in the utterance. If the value of a ranking function of an utterance is more than the threshhold, and if a cluster Ci is the largest cluster and Cj is the second

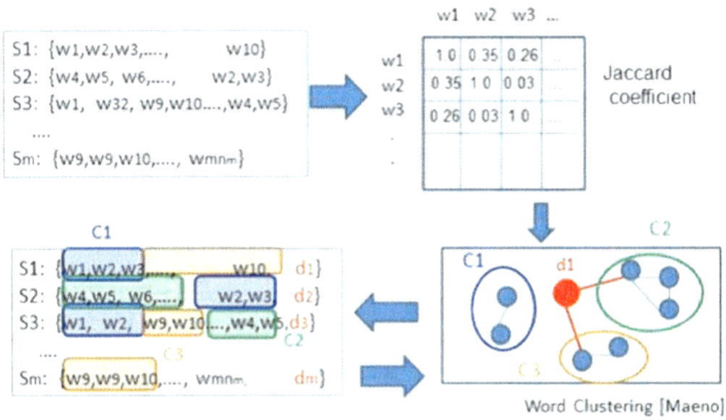

Fig. 4. Word Clustering Method and Dummy Nodes

largest cluster, then from a dummy node to the representative nodes of the Ci and Cj, we make links. In the example in Fig. 4, a dummy node d1 combines C2 and C3 which means that C2 and C3 are large clusters in the utterance S1.

By interpreting the role of dummy nodes, we can extract various kinds of information. For example, some dummy nodes correspond to utterances where one topic is spoken by reffering to another topic. Another dummy node corresponds to utterances where one topic is spoken, wanting to change the topic to another one.

3.2 Temporal Word Clustering Method

When the discussion record is small in size, the original word clustering method works well. However, when the size becomes bigger, precision of word clustering decreases because the role of each word may change during the discussion and because a small cluster may be absorbed into a bigger one. To cope with the problems, we devised the temporal word clustering method. This method divides the discussion records into several parts at the points where the topic changed a lot as follows.

At first, for the total discussion record, the original word clustering method is applied. Then, we count the number of words for each cluster in chronological order, and make a histogram in which each cluster is represented by a line. In the histogram, we find time points where two lines cross, and at these time points, the record is divided into several sub records (discussion periods). Then, for each sub record, the above process is applied hierarchically. As a consequence, if a discussion record is divided into N sub records, the output of the temporal word clustering becomes N clustering graphs (Fig.5). As we divide the discussion record leaving some overlaps, dummy nodes in these overlaps may combine two clusters which belong to adjacent word clustering graphs. These dummy nodes correspond to utterances which caused the change of topics.

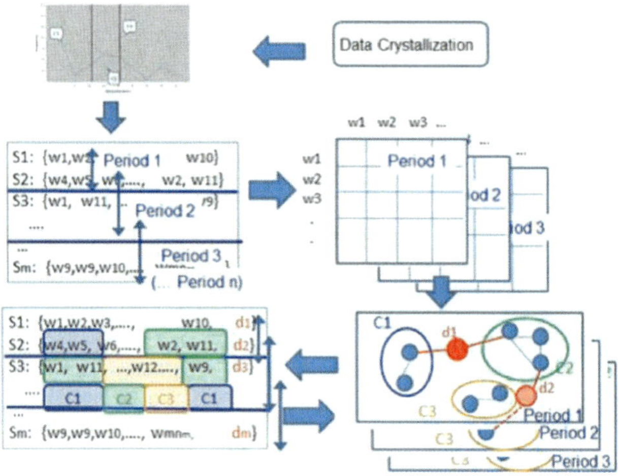

Fig. 5. Temporal Word Clustering Method

3.3 Multimodal Temporal Word Clustering

Dummy nodes correspond to utterances which include more than one topic. By observing these topics, we can estimate important changes in topics. However, when there are utterances that include more than one topic, their interpretation is not easy.

Sometimes, discussion records are given in the form of a movie file, and we can observe facial expressions, actions such as crossing arms, nodding, hand waving, falling forward and so forth. Such nonverbal information is useful to estimate the emotional state of individuals while they are speaking. Therefore, by combining a temporal word clustering method and multi modal information, we can extract more detail information which supports the interpretation of dummy nodes.

Nonverbal information in the discussion records can be analyzed as follows. At first, we observe the movie file and extract nonverbal information listed in Table 1. The extracted information is tagged according to each utterance by using iCorpus Studio [4] , and is saved in an XML format as shown below.

```
<Utterance id="15" Speaker="Yamada" Issue="F3"
   Head="Tilt" Body="Straight" Arms="Crossed">
    Contract dissolution will not be granted
</Utterance>
```

Then, each utterance Si is represented as follows.

$$Si = \{\ w1,w2,\cdots,wn;\ a1,\ a2,\cdots,\ am;\ di\ \}$$

Here, "w1, w2,···, wn" are words which appeared in Si, "a1,a2,···,am" are the name of the speaker and nonverbal information which appeared in Si, and "di" is a dummy node.

Table 1. Type of nonverbal information

Body Part	Type
Head	Straight, tilting to right, tilting to left, upward, downward, nodding, sticking out, leaning
Body	Straight, tilting forward, tilting backward, tilting to right, tilting to left, swinging
Arms	Crossed, putting the hands forward (vertically), putting the hands forward (horizontally), touching the chin, hiding the mouth, touching the head

We show an example of a TV discussion program, "Asa made nama TV." In this program, 14 people joined and discussed about 6 problems with the Japanese Government such as the manifesto of Japanese Democratic Party, Okinawa's military bases, Defense of Japan, Economical stimulus policy, Government's Finance, and so on. Participants are composed of economists, journalists, militarists, psychologists and representatives.

During the 4-hour discussion, several actions were observed as seen in Table 2. In this table, from Head1 to Head3 correspond to positions of the head such as 'downward', 'straight' and 'nodding', respectively. From Body1 to Body5 mean positions of the body such as 'tilting right', 'tilting backward', 'tilting left', 'tilting forward' and 'swinging', respectively. From Arms1 to Arms6 means movement of arms such as 'touching chin','putting hand forward', 'putting hand aside', 'putting hand vertically', 'crossing arms' and 'crossing fingers', respectively. In this table, we can observe several actions of speakers. For example, Mr. Yamagiwa and Ms. Kayama showed several actions while they were speaking. On the contrary, Mr. Morimoto and Mr. Uesugi showed little actions. If these actions are affected by certain emotional status, then by observing the relation between these actions and topics, we can estimate the role of each one's utterance to a greater degree.

Table 2. Actions during discussion

Speaker	# speak	Head1	Head2	Head3	Body1	Body2	Body3	Body4	Body5	Arm1	Arm2	Arm3	Arm4	Arm5	Arm6
Tahara	433	0	3	0	0	0	0	9	1	10	0	3	9	1	0
Itokazu	33	0	1	0	0	0	0	1	0	0	0	2	0	0	0
Uesugi	63	0	0	0	0	0	0	4	0	0	0	0	0	0	0
Ohtsuka	99	0	2	0	1	0	0	1	0	0	1	2	7	18	0
Katsuma	33	0	9	0	0	0	0	1	0	1	0	0	0	0	0
Kayama	13	0	2	0	0	0	0	1	0	1	1	3	0	0	0
Kawauchi	106	2	1	6	0	0	0	2	0	0	3	13	0	2	1
Motegi	77	1	1	2	0	0	0	4	0	1	0	10	0	0	0
Morimoto	51	0	0	0	0	0	0	0	0	0	0	3	0	0	0
Yamagiwa	61	0	5	0	0	2	1	24	0	0	0	19	0	0	17
Yoshizaki	20	0	0	0	0	0	0	1	0	0	0	0	0	2	0
Takahashi	58	0	0	1	0	0	1	1	0	0	2	3	0	2	0
Takano	58	4	1	0	0	1	0	4	0	0	1	4	1	0	0
Koike	78	1	0	1	0	4	0	3	0	8	0	5	3	0	0

There are two ways to analyze the nonverbal information. Method 1 is to consider a multimodal label as a word, and apply the temporal word clustering method to the following basket.

$$si = \{ w1, w2, \cdots, wn, a1, a2, \cdots, am \}$$

In this method, each nonverbal label belongs to one of the clusters, which means that such a label is related to the cluster the most. For example, if the nonverbal label is the speaker's name, it means such a topic that he talks about most. If the nonverbal label is an action such as "crossing hands", it means that "crossing hands" are observed during the discussing such a topic.

Method 2 is that we treat multimodal labels as attributes of the dummy node as follows.

$$di(a1, a2, \cdots, am)$$

In this method, the result of the temporal word clustering with nonverbal information as the same as that of temporal word clustering without nonverbal information. However, to each dummy node, several other forms of information such as the speaker's name, action labels which is observed while speaking and so on are attached. This information is then used to interpret the meaning of the dummy node.

Fig. 6(a)-(f) shows the results of Method 1. There are 6 figures, because the discussion is divided into 6 periods. In these figures, several kinds of information such as names of speaker, actions, and roles of utterance appear as nodes. As Mr. Tawara is the chair person, his name appears in various clusters throughout these figure, and there are a lot of dummy nodes from Mr. Tawara connected to other clusters. To these dummy nodes, the roles of the utterance are attached. According to this information, most of what Mr.Tawara's speaks are YES/NO questions, open ended questions, agreements and so forth.

Several action labels appear on the upper left cluster which corresponds to the topic of "violating manifesto." This topic is a very general one and the other 5 topics are closely related to this topic. Therefore, even if the other topic is the on-point one, the topic "manifesto" is often referred from this topic. In Period 1, "Transferring Okinawa's military bases" is discussed. While 4 participants mainly talked about this topic, most other participants talked "violating manifesto." Most nonverbal labels occurred in these two topics. In Period 4, "financial problem" is mainly discussed. In this period, most nonverbal labels appear in "violating manifesto" topic because this topic tends to be the most exciting one.

Now we will show a more concrete example. Fig.3.3 shows the result of Method2. Six clusters corresponds to six topics such as "keeping promises described in manifesto", "transferring military bases in Okinawa to other prefectures", "importance of the relationship with the United States for national defense", "financial deficit of the Government and consumption tax", "stimulating the economy" and "fraction of DPJ." In this figure, several dummy nodes appear. We focus on two dummy nodes 304 and 310 as seen in Fig. 3.3. To these dummy nodes, multimodal label "tilting forward" is attached, which means these utterances were made emphatically. In Fig. 3.3, utterances around these nodes are shown. Actually, these utterances played an important role to bring about change.

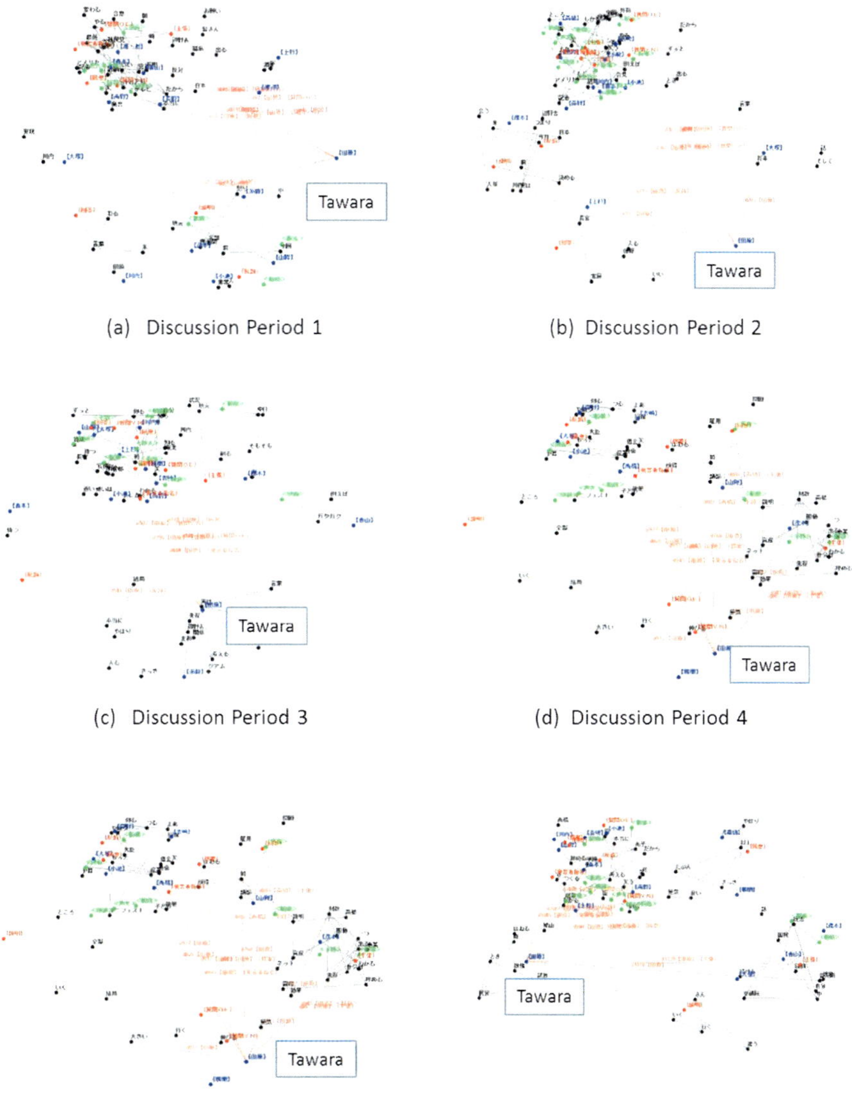

(a) Discussion Period 1

(b) Discussion Period 2

(c) Discussion Period 3

(d) Discussion Period 4

(e) Discussion Period 5

(f) Discussion Period 6

Fig. 6. Word Clustering Graphs

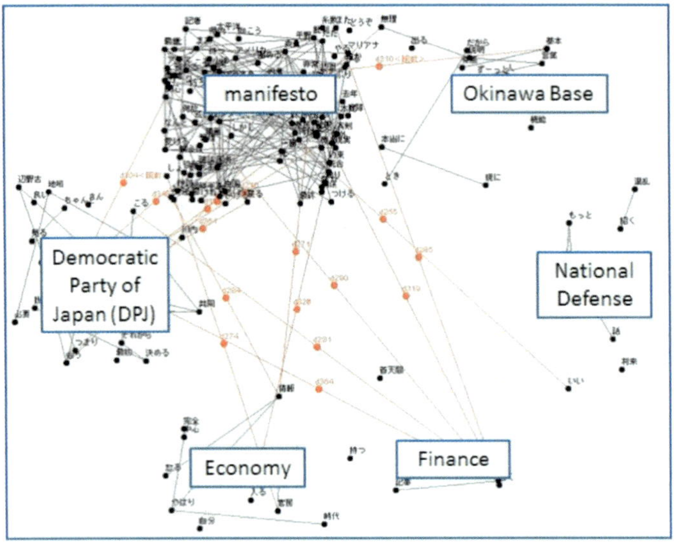

Fig. 7. Result of Multimodal Temporal Word Clustering

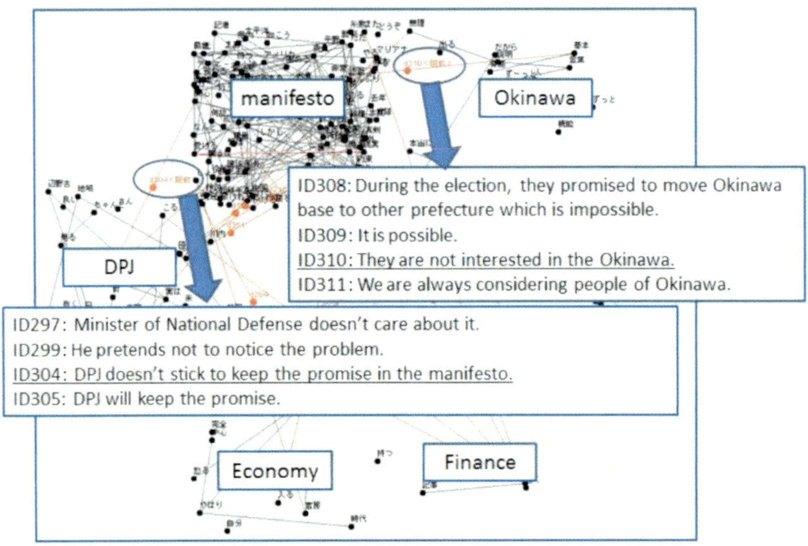

Fig. 8. Dummy Nodes 304 and 310

4 Conclusion

Here we have shown a novel method for analyzing discussion records. This method uses both logical analysis of the discussion subject and statistical analysis (a temporal word clustering method) of discussion records. By integrating these two methods, we can evaluate discussion moderation skills and the analyzed data that are reused in other discussion where the subject is the same. Furthermore, we showed a method for nonverbal information by extending the temporal word clustering method.

Acknowledgement. This research was supported by AFOSR/AOARD grant no. 114061. I thank AFOSR/AOARD for their support. Also I thank Professor Yukio Ohsawa (The University of Tokyo) for his technical advice about the word clustering method.

References

[Reed 04] Reed, C., Rowe, G.: Araucaria: Software for argument analysis, diagramming and representation. International Journal on Artificial Intelligence Tools 13(4), 961–979 (2004)

[Aleven 97] Aleven, V., Ashley, K.D.: Teaching CaseBased Argumentation through a Model and Examples: Empirical Evaluation of an Intelligent Learning Environment. In: Proceedings of AIED 1997 World Conference, pp. 87–94. IOS Press (1997)

[Ohsawa 06] Ohsawa, Y. (ed.): Chance Discovery in Real World Decision Making. Springer (2006)

[Maeno 06] Maeno, Y., et al.: Crystallization highlighting hidden leaders. In: Proc. IPMU (2006)

[Maeno 09] Maeno, Y., Nitta, K., Ohsawa, Y.: Reflection of the agreement quality in mediation. In: Proc. 3rd International Workshop on Juris -Informatics (Jurisin 2009), pp. 73–82 (2009)

[Nitta 09] Nitta, K., et al.: Scenario Extraction Support System Using Word Clustering and Data Crystallization. In: Proc. 3rd International Workshop on Juris -Informatics (Jurisin 2009), pp. 95–106 (2009)

[Sato 11] Sato, T., et al.: Deliberation Process Support System for Citizen Judge Trial Based on Structure of Factors. In: Proc. 5th International Workshop on Juris -Informatics (Jurisin 2011) (2011)

[Omoto 11] Omoto, K., et al.: Generation of gesture of embodies conversational agent by using a case base. In: Proc. Human Agent Interaction Symposium (2011)

Framework of Early Adopters Roaming among Tribes for Discovering Innovative Creation

Mu-Hua Lin[1], Chao-Fu Hong[2], Hsiao-Fang Yang[1], and Ai-Ling Wang[3]

[1] Management Information Systems, National Chengchi University, Taiwan
[2] Department of Information Management, Aletheia University, Taiwan
[3] Department of English, Tamkang University, Taiwan
95356503@nccu.edu.tw, au4076@au.edu.tw, hfyang.wang@gmail.com, wanga@mail.tku.edu.tw

Abstract. Our research pointed out that traditional Data Mining (DM) only construct clusters of consuming products and does not use the relation of consuming products to explore the relation between consumers to develop consuming tribes. So, the social science is not easily for observing any social behavior from result of DM. Our model tend to break through this neck bottle to transform the clusters of mostly bought product to consuming tribes and to find out weak ties, who roams over tribes. And this social phenomenon will bring important information for us, what innovative use of new products will blow a tornado in future.

Keywords: innovative creation, tribe, social phenomenon, roaming among tribes.

1 Introduction

In his experimental results of innovation diffusion, Rogers [16] has pointed out that early adopters tend to accept innovative products more easily and more readily. On the other hand, early adopters are able to develop creative ways to use innovative products. Early adopters may act as lead users to affect the early majority to accept the product [19,10]. Creating innovative ways of using a product can be one way for early adopters to convince the early majority to accept the product. This can be a product diffusion model. In the process of diffusion, people are building their social blocks and forming tribes or neo-tribes. This can be found in social groups formed because of preference particular products or brands or having similar life experiences or ideas. Their meaningful social relationships, as Cova (1997) [3] has argued that connect them, are more important than things (brand, products, experiences, ideas).

In investigating the market, data mining is often used to analyze the consuming data and to find out consumers' behavioral patterns. However, even though the theory of Chance Discovery (CD) can generate an associative network of consuming products, researchers can hardly figure out what kind of social relationships might facilitate the formation of social tribes, and they might have

Y. Ohsawa & A. Abe (Eds.): Advances in Chance Discovery, SCI 423, pp. 91–106.
springerlink.com

problems understanding how innovate products are diffused in a consuming network, like the way Innovation Diffusion Model (IDM) does. Therefore, in this research we not only try to explore the associative products network, but also try to discover what consuming social relationship maintained by early adopters to construct tribes.

The following story told by Granovetter [4] illustrates how social networks work: a labor was looking for a career change, and he and his close friends (strong tie) shared the same job information. One day, friends from other groups (tribes) came to see him and bring him information about new jobs. This information is very important for him in job hunting. This social phenomenon means that early adopters roam among the tribes, to get a chance for integrating differential use innovations of tribes to create an innovative use. So, the scenario demonstrates how, in Rogers's model [16], early adopters may accept new products and may create innovate ways of using the products to convince the early majority to accept the new products.

Summing up the above mentioned, we decided on two focuses of our research study: one is to identify appropriate products to be diffused in the tribes, and the other one is to explore the phenomenon as to how innovation is created. For example, weak tie or lead users who belong to different tribes may reveal innovative ways of using a product. In this study, we focused on exploring the social phenomenon describing the creation of innovation and tried to develop a method to discover weak tie or lead users and their innovate use of a product. After the invention of Web 2.0, it is much easier for people to post on the Internet and for researchers to gather textual data related to innovative products in virtual communities.

2 Literature Review

When new products are introduced to the market, according to Roger's IDM only early adopters will purchase new products and create innovative ways of using them. Furthermore, from a marketing perspective the more tribes are merged, the larger market is built. Therefore, we may connect tribes shared by early adopters to build larger social networks. In this section, we will review relevant literature to demonstrate our research framework.

2.1 Social Network and Social Influence

A social network is a graph of relationships and interactions within a group of individuals, which often plays a fundamental role as a medium for spread of information, ideas, and influence among its members [1]. A Web-based social network provides various ways for social interactions such as a chat room, a discussion forum and a weblog. Participants can interact, exchange opinions, and compare experiences with others. In E-commerce, social networks emerge since many websites help a consumer's final purchase decision by sharing reviews written by previous customers and evaluated by potential customers. In addition, as Bickart & Schindler [2] pointed out, there are two kinds of social influence in the adoption of an innovative product: normative social influence (or subjective norms) and informational social influence. Normative social influence creates social pressure for people to adopt a product or a service. Informational social influence is a learning process through which people observe the innovative use of a product by early adopters in their social network and decide to buy the innovative product.

Other marketing researchers suggest viral marketing is that targets a few influential consumers initially to trigger a cascade of influence through a social network in which friends will voluntarily share their experiences or recommend the product to other friends [1]. However, when looked at from our tribal perspective, human social life are in a multiple social groupings that we all participate in, consciously or unconsciously, through the course of our daily life. These tribes or little masses (known as neo-tribes) are fundamental to our experience of life in general, such as consumption of cultural resources circulated through markets (brands, leisure experiences and so on). This means extracting key early adopters and their innovative ways of using a new product are very important for triggering social influence. In the next section we will discuss what early adopters are and how to identify him.

2.2 The Innovative Use of Early Adopters

In this section we are going to discuss the innovative use of a product created by early adopters. Firstly, [14] defined innovative use as "the degree to which an individual is relatively earlier in adopting an innovation than other members of his social system." This definition of innovation is limited to a purchasing context. [6] extends the concept of innovation to two other categories - use innovation and vicarious innovation. A consumer is use innovative, "when he or she uses a previously adopted product to solve a novel consumption problem." "Vicarious innovativeness refers to active information search and interest in a new product, but not actual purchase." Hence, innovation can manifest itself from the purchase context to the usage context.

Therefore, creative consumers such as early adopters may possess special skills and abilities required for using the product in a wide variety of ways [13]. Besides, if a product is quite innovative and discontinuous, only creative consumers may discover additional ways of using the product [14,18]. Therefore, creativity is not only a significant value of variety use innovation and purchase innovation, but can also add to the values of early adopters. In order to test early adopters how affect early majority to accept innovative products, we should first of all to identify the early adopters and extract the creative value from them.

However, the research results found by [15,19,10] show that identifying early adopters in the traditional market is very difficult and it costs money. Fortunately, the invention of web 2.0 makes the Internet no longer a monopoly to enterprises. The consumers can post their messages on the Internet. In the same token, traditional ways of transmitting message by word of mouth can gradually be applied to the Internet world. For example, from online discussions (e.g., Internet forums or bulletin boards) or customer-generated online information (e.g., corporate webpages) [2], consumers who gathered information from online discussions reported greater interest in the product topic than did those consumers who acquired information from the marketer generated sources.

As mentioned above, researchers may collect data from the Internet the chance to easily discover early adopters and some innovations they create and can find them and extract the purchase innovation and use innovation from them. Besides, the cost of searching data on the Internet is much lower than that of traditional statistical sampling method. The methods to extract purchase innovations and use innovations from the early adopters will be discussed in the following section.

2.3 Linkages and Clusters

In this section, we are going to examine two information techniques relevant to linking clusters. One is KeyGraph algorithm proposed by [12]. KeyGraph algorithm resembles Apriori algorithm, which calculates the frequency of the node in the shopping cart and the co-occurrence of two nodes, in which clusters are expected to emerge. The other one is the keyvalue derived from all the nodes connecting to all clusters. From the clusters, researchers may find nodes (products) with high key values and low frequency, which are called chance nodes. Then the researcher may integrate the hints given by chance nodes and clusters to build a chance scenario, which is called chance discovery. However, the result acquired by using this method is too complex, and it is not easy for researchers to conduct the analysis [11], and it lacks the information required to understand the transforming relationship between the chance product and early adopters to support IDM.

[7] proposes Qualitative Chance Discovery (QCD), in which he adopted the co-occurrence clusters to analyze research issues on segmentation aiming at addressing the above-mentioned problems. Several steps were taken to achieve the goal. First, the researcher is to look for interesting theme terms and theme attributes terms from the initial association graph. These theme terms and related attribute terms are used to extract relevant documentary data. Then, the researcher focuses on the theme which holds the characteristics of interestingness and single value. By removing irrelevant terms, those terms that show simplicity and readability will emerge. Repeat the steps mentioned above until all the theme structures with single values are extracted. Finally, researchers, based on the chance scenario, integrate the related single-value theme structures. Only if the node in the structure repeatedly, a value scenario with multiple structures can be created. This is similar to a dichotomy association network. This kind of multi-value theme structure with more value-focused orientation will help researchers find out the short-route linking nodes existing between one single value and another single value. This is similar to chance nodes, which are the starting points for developing future scenario. This method still lacks the information required to understand the relationship between the chance product and the early adopters to support IDM.

After all the discussions above, we may have a rough idea as to how innovative products are diffused in the social network. First the social network has to be constructed. Second, how products are diffused through social in has to be analyzed. However, previously-mentioned methods have problems helping us construct the social network. Therefore, in the research, we attempt to develop a new method to solve the problem. In the social network we construct, we quantify each customer's social position to identify key earlier adopters and their various use innovations.

3 Methodology

In [15] innovation diffusion model, when early adopters evaluate innovative products, they tend to take products properties and personal consuming preference into consideration. Also, since the number of early adopters is relatively smaller, they should not only equip themselves within a social network but also possess the traits of lead users, who can create innovative use of a product as a need-forecasting for marketing research [19]. This is a phenomenon that early adopters are challenged to make use of a peer-to-peer contagion effect at an early stage of the introduction of a new product. It can be visualized as a node may have a certain degree of effects on its neighbors. However, when looked at from the perspective of homophily, dyadic similarities between two nodes may reveal similar patterns that do not show direct casual influence and that cannot happen in an early stage. Based on the assumption, this study try to investigate how a lead consumer may have effects on other consumers in a consuming tribe. In addition, if he is a member of different tribes, he may serve as a weak tie, bridging and affecting members of the entire tribes to create an innovative market. Therefore, this study aims to propose a model to identify lead customers who belong to different tribes and who are said to be more influential in affecting other customers' decision. The research flowchart is shown as follows.

3.1 Human-Centered Computing System

In Rogers' IDM [16], early adopters create innovative ways of using the products and pass on these novel ways to the early majority to motivate them to accept the products. Here, the uses of innovative products are conceptualized as a consumer's receptivity/attraction to and creativity with using innovative products in new ways [10]. To return to Rogers' IDM, if the early majority could not obtain or accept the uses of the innovative products of the early adopters, the chasm will be existed between them [9]. This means that the uses of innovative products are the key factors of Social Influence (SI) for influencing the majority to accept the innovative products. In this study we used grounded theory (GT) [17] and employed text-mining method to develop a Human-Centered Computing System (HCCS). The following steps were followed in the entire research process.

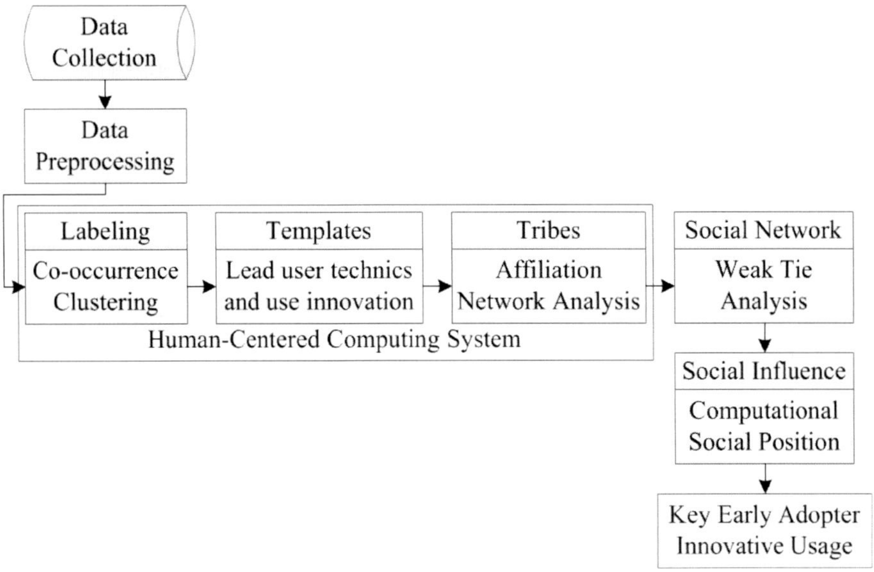

Fig. 1. A research flowchart for extracting key early adopter

Phase 1: Preparation for data and labeling process

To create innovation diffusion scenario based on social influence, the researcher first collects data of the innovative uses of a product to conduct a preprocessing on the data and then combine GT with text-mining to process the data. The detailed process is listed as follows: H labels indicate the step to be done by researchers, and C labels indicate the step to be done by the system.

Step 1: Data preprocess

1-1-1H) The researcher defines the domain and relevant keywords he/she intends to study.
1-1-1C) The researcher sifts out the data which correspond to keywords from the Internet.
1-1-2H) Based on his/her domain knowledge, the researcher interprets the texts, and at the same time, segments texts into words, removes useless words, and marks meaningful words as conceptual labels.

Step 2: Labeling process: words co-occurrence analysis

1-2-1C) Use equation (1) to calculate the association values of all words as below:

N is all words
i=1 to N-1
j=i+1 to N

$$assoc(w_i, w_j) = \sum_{S \in allD} min(| w_i |_s, | w_j |_s) \quad (1)$$

where s represents the co-occurrence of words in the sentence, and D represents all textual data.

1-2-2C) To visualize the analysis result, the system contributes to an associative diagram drawn by the co-occurrence words.

1-2-1H) The researcher identifies keywords as concepts and the clusters as categories derived from the co-occurrence association diagram, and gives the clusters' label, such as $topic_1$, $topic_2$, and so on, which helps the researcher to preliminarily realize the various theme values presented in the data.

Phase 2: Construct the early adopters' tribe (social network) by template of lead users

Based on the analysis done in phase 1, which the researcher discovers various types of uses for the innovative products, clusters are derived from early adopters, and the different types of using innovative products also are used to assign early adopters to different clusters. Therefore, in the same cluster of innovative use, early adopters have similar techniques and use innovations. If technique the same or use innovation connects two early adopters, the connection serves as the linkages between them. Then, all linkages of technics or use innovations are supported to build the use innovative affiliation network of early adopters. So, the affiliation network supports the system to construct the early adopters' social network as the tribe. The detail processes are as following:

Step 1: Use the template of lead users to generate the use clusters of innovative products.

In this step, the leader user, who can lead users' "leading edge status" and their in-house technical capabilities not only improve the original system's efficacy (OPACs), but also enable many innovative users freely sharing their innovations with others [1], and they can create use innovation as a need-forecasting for marketing research [19]. So, the researcher uses "the technical capabilities and use innovation" as the template of lead users to extracts the documents and search out the useful sentences data to create a use clusters of innovative product. The process is illustrated as follows.

2-1-1H) The researcher needs to decide what use innovative topics $(topic_1, topic_2, \cdots)$ are.

2-1-1C) the system uses the clusters' label $(topic1, topic2, \cdots)$ to pick up the documents. In this step we want remove the irrelevant article, so if the frequency ranking of topic's word is not in the top 50%, then the article is removed.

i=first topic to last topic

$$varid\ document\ set(topic_i) = \sum_{D \in allD} if(\{w_1, w_2, \ldots\}_D \cap \{topic_i\}) \neq \phi \quad (2)$$

2-1-2C) Use equation (1) to calculate the association value of all words in the set of valid document, and then create topic co-occurrence graph.

Step 2: Construct tribes using social affiliation networks

In the same topic cluster, the techniques and use innovations of early adopters are similar and the system can help researchers build the use innovative affiliation network of early adopters and construct early adopters' social network.

2-2-1H) Researchers examine the topic graph and the template of lead users definition [8,14] to identify the value words $(vw_1, vw_2, \cdots)$.
2-2-1C) The system use topic's label $(topic_1, topic_2, \cdots)$ and value words set $(vw_1, vw_2, \cdots)$ to dig out valid sentence sets.

i=first topic to last topic
varid sentence set(topic) =

$$\sum_{S \in valid\ document\ set(topic_i)} if(\{w_1, w_2, \ldots\}_s \cup \{vw_1, vw_2, \ldots\}) \neq \phi \quad (3)$$

2-2-2C) Use equation (1) to calculate the association value of all words in a set of valid sentence, and then create a co-occurrence diagram.
2-2-3C) Construct tribes. Based on the bipartite network analysis, e.g. in the case of movies connecting to actors, researchers can build actor networks. Here the value words are the connecting relationship.

MN is all early adopters
VN is all value words
i=1 to VN
j=1 to MN-1
    k=j+1 to MN
$assoc(mb_j, mb_k) =$

$$\sum_{S \in valid\ sentence\ set\ of\ all\ topic} min(assoc(mb_j, vw_i), assoc(mb_k, vw_i)) \quad (4)$$

Phase 3: Based on early adopters' network to extract key early adopters

In this phase, the system, based on central network of social network analysis, may figure out the social position of every early adopters and identify key early adopters. In this phase, the centrality of social network analysis is used to measure how an early adopter is close to the center of network and how important the early adopters is [8], [5].

3-1-1C) Computing the social position of early adopters (SC).

i=1 to MN-1

j=i+1 to MN

$$APL_G = min(APL_G + \frac{1}{MN(MN-1)} \times d(mb_i, mb_j)) \tag{5}$$

k=1 to MN

$$SC_{mb_k} = APL_G - APL_{G-mb_k}$$

When a social network includes the early adopters mb_k, APL_G is average distance among all early adopters, and when the early adopters mbk is removed from the social network, the average distance between all early adopters is APL_{G-mb_k}. Therefore, the smaller the value in equation (5) is, the more important the early adopters in the social network.

4 Case Study

In 2007, ASUSTeK Computer Incorporated (ASUS) and Intel Corporation (Intel) collaboratively introduced to the market a less-than-10-inch Netbook. They aimed at providing more people with inexpensive computers. However, the follow-up investigations done by major industries indicated that netbooks have increased their effectiveness and so have their prices. This was not really the initial purposes of inventing netbooks. Thus, industries once again refer to netbooks as being light, thin, small, and easy to carry as consumers' second notebook. However, how to make consumers realize the new use value created by netbooks is an issue for industries to address.

4.1 Data Resources

After browsing on the Internet the messages about circulation in the market circulation messages in the market, we found out that there is a time gap in transmitting speed on the Internet between introduction of products and information on how users use them. This study focuses on the target innovative product: ASUS EeePC901. Therefore, we collected six months after it was introduced to the market to conduct the analysis. The dates include posted messages between October 1, 2008 and October 31, 2008. Using Google blogs (`http://blogsearch.google.com/blogsearch`) and giving key terms, netbook and thought, to search for the data, we obtained 63 related data from blogs articles.

In addition, [14] define purchase innovativeness and [6] extends the concept of innovativeness to two other categories — use innovativeness and vicarious innovativeness. At the same time, from [10] experiment, we recognized that lead users' "leading edge status" and their in-house technical capabilities not only improve the original system's efficacy (OPACs), but also allow many innovative users freely share their innovations with others. This is may explain that purchase innovativeness and use innovativeness tend to affect the majority's acceptance of the new products. Therefore, after carefully reading the collected data, we removed articles irrelevant to use innovativeness. Finally, we kept 7 related articles for further value extraction.

4.2 Labeling

The researchers based on their knowledge and followed the phase one of human-machine interaction of HCCS, as shown in Fig. 1, did labeling and clustering process. First, the researchers must read documents carefully to dig out meaningful words. Then, the researchers read carefully those meaningful words for some concepts lying behind those clusters of meaningful terms. Those emerging concepts then named. At the same time, those words with similar meaning and concepts were categorized as concepts. Next, these meaningful words will be labeled with concept words. Then, the researchers focus on new products' functions used by early adopters and early adopters' use of new products which generate new service value, namely, new products' functions and new products' new use value. Then, through the calculation of words frequency of co-occurrence value, the researchers visualized the association results as shown on Fig. 2. We did this just like the concept classification done by the early adopters.

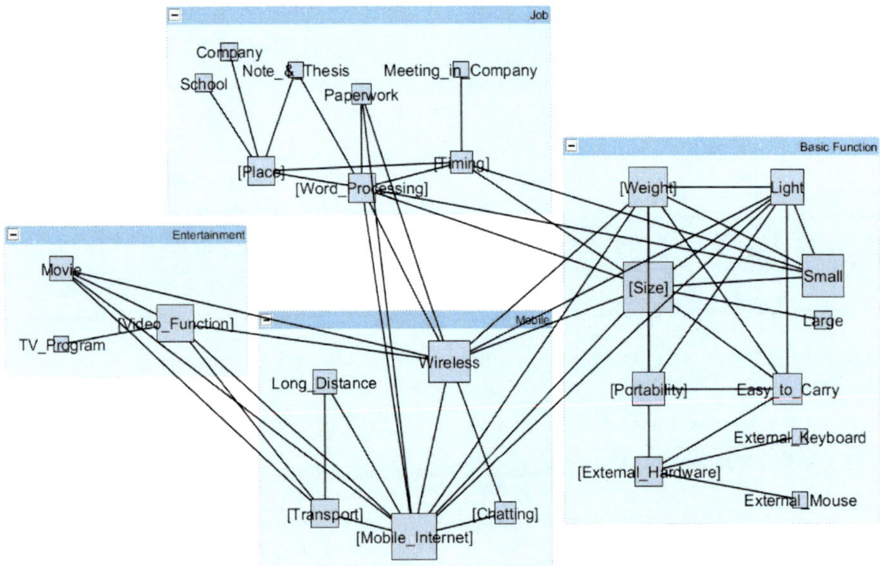

Fig. 2. Open coding

4.3 Template: Explore How Netbooks Were Used by Early Adopters

According to section 4.2, most of the use innovations inspired the researchers to cluster the categories and only four categories were discovered in Fig. 2. In addition, key words relevant to use innovation topic were selected by the researchers from the topical clusters. For instance, *Entertainment, Mobile Internet*, and *Job* were used to dig out useful articles. These selected articles stories were used to build each topic's associative map.

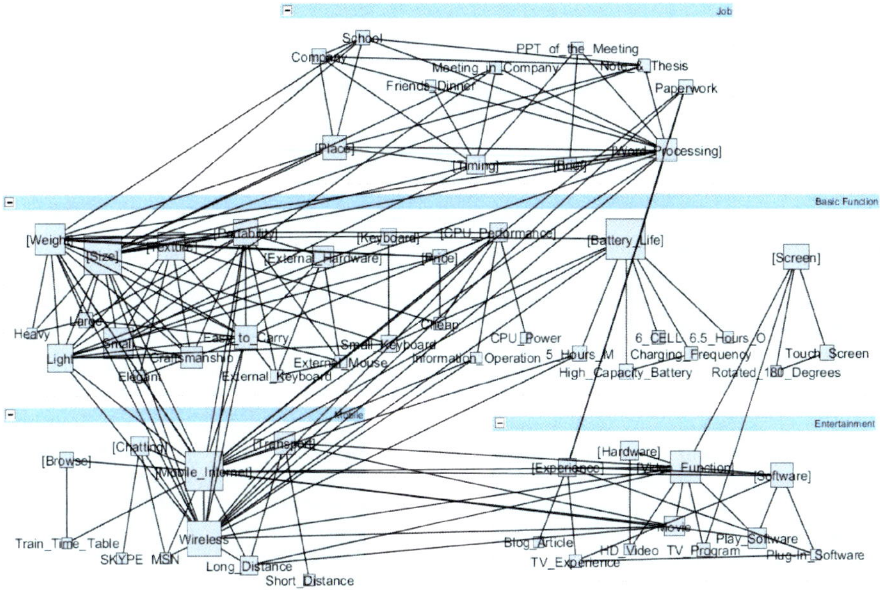

Fig. 3. Document base mobile internet associative map

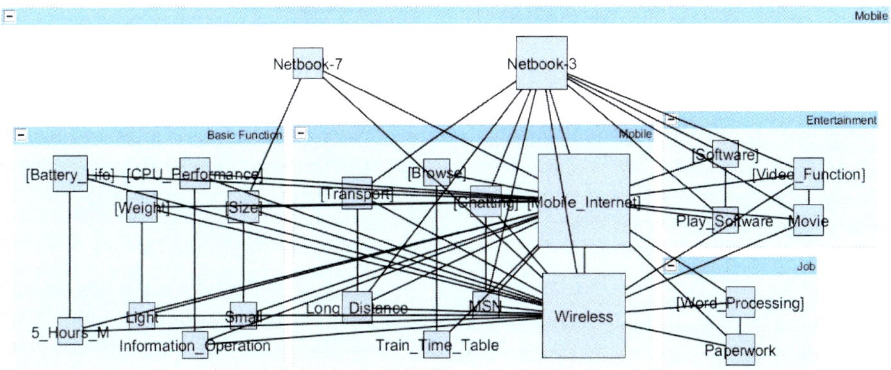

Fig. 4. Mobile internet tribe

Then, researchers used their knowledge to label the way netbooks were used be early adopters with perceived ease-of-use (PEU) and perceived usefulness (PU) of TAM. First, the researchers identified some PEU characteristics of netbooks, e.g. being light, small, wireless, and its long-lasting battery. Second, the researchers also found that consumers have innovative ideas to use netbooks. For example, netbooks provide them with browser, MSN, BBS and Skype to communicate with their friends. It is easy for customers to carry a netbook to anywhere and

Table 1. Template's Definition of Lead users

Construct	Extracted characteristics from early adopters / Definition of operational variables
PEU-Physic	Size, external hardware, thickness, weight, hardware, battery life, price, texture, color
	Large, small, external mouse, external monitor, external keyboard, reader machine, thin, heavy, light, d sub, led backlight, cannot be projected, home theater, sound terrible, not synchronized, 4 cell, 6 cell, charging frequency, 5.75 hours (o), 6.5 hours (o), high capacity battery, 5 hours (m), cheap, expensive, elegant, craftsmanship, white, black
PEU-Operating	CPU performance, screen, storage, keyboard, touchpad, portability
	CPU Power, middle preference, quick power, information operation, high resolution, rotated 180 degrees, acceptable quality, widescreen, touch screen, sensitivity, 12g, 16g, SSD HD, external HD, limited number of write, small keyboard, multi-touchpad, touchpad problem, not easy to carry, easy to carry
PU-Entertainment	Experience, word input, software, video function, game, picture/photo
	Reading experience, travel experience, blog article, TV experience, key-in, plug-in software, play software, youtube, HD video, TV program, movie, entertainment, game, digital photo
PU-Mobile	Transport, mobile internet, chatting, communication, browse
	Long distance, short distance, GPRS, ipod transfer, wireless, BBS, MSN, SKYPE, calls when traveling abroad, webcam, microphone, NBA website, wiki, train time table
PU-Job	Word processing, place, timing, brief
	Paperwork, note & thesis, record of the meeting, company, meeting in company, friends dinner, cafe, school, recording in class, PPT of the meeting

to do word processing or to play video or audio. Therefore, the researchers easily identified three kinds of templates: Mobile Internet, Entertainment and Job. The results are shown in Table 1.

4.4 Tribes

The data were collected and categorized by the researchers, following Step 2 in Phase 2. Basically, the researchers noticed that customers can easily carry a netbook to anywhere at anytime to have immediate access to the Internet because of its being light and small in size.

Especially, consumers agree that they share immediate information with others in written form, photos, and meeting in company or school, as shown in the job's template. By the same token, in their daily life, customers may use netbooks to share with their good friends the movies they have watched, the nice travel they had, and the products they bought. They also receive the

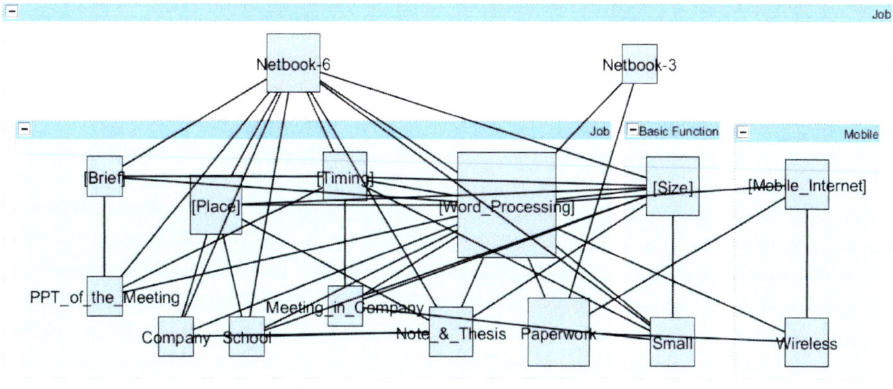

Fig. 5. Job use tribe

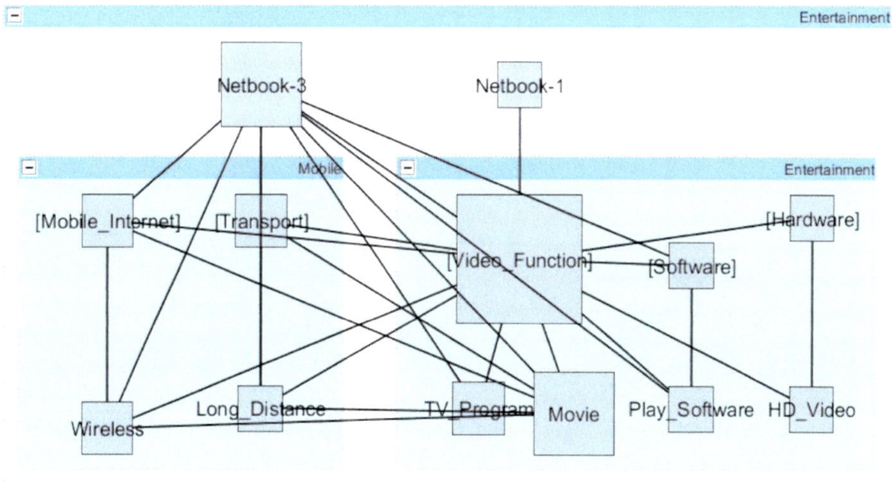

Fig. 6. Entertainment use tribe

information shared by their good friends, as shown in the mobile internet template. Finally, they can also enjoy movie, music and game as the entertainment template has shown. These templates help the researchers not only explore how use innovations might motivate customers to accept the products, but also find early adopters' use innovations to dig out the sentences relevant to the template. Those data allow the researchers to construct a template's affiliation network, and by doing bipartite network analysis, the template's affiliation network is transformed to a human-based consuming network known as tribe. Therefore, in this step the researcher can construct job use tribe, mobile internet use tribe and entertainment use tribe, as shown in Fig. 5 and 6 .

4.5 Social Influence

Weak tie like short cuts between different clusters (tribes), which can help the researchers to discover the potential innovation. Therefore, in this step, a number of different tribes were synthesized by the researchers, i.e. we integrated different tribes for emerging weak ties. The researchers combined job use tribe, mobile internet use tribe and entertainment use tribe, and the result is shown in Fig. 7 and 8. From Fig. 7 the researchers get an early adopter, who like weak tie, roams over different tribes. In Fig. 8, he tries to integrate the action of leisure (mobile internet tribe and entertainment tribe) tribe and the action of job tribe to create a new way of using netbooks (life), for example in travel he enjoys chatting or sharing the photo of travel with his friends, watching movies or TV, and reading.

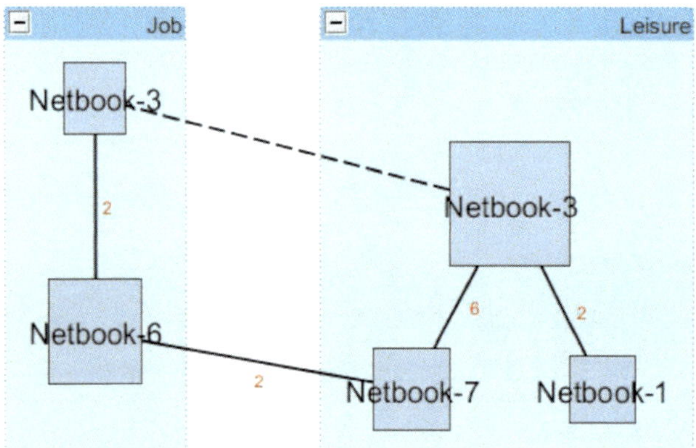

Fig. 7. Synthesis tribes to emerge weak tie

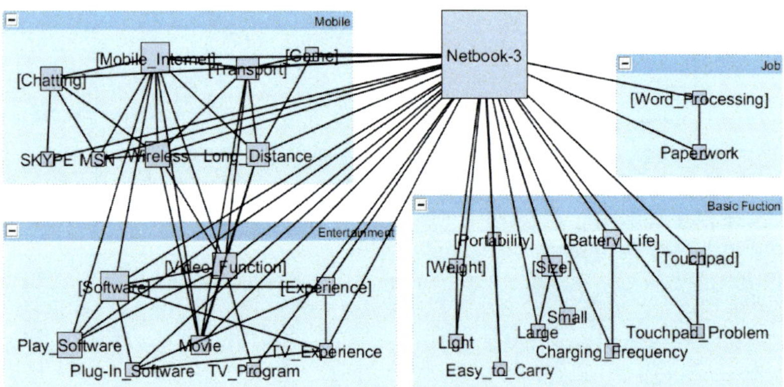

Fig. 8. Roams around the tribes

After doing the qualitative analysis, the system calculates the social position of early adopters based on the human-based affiliation network. Here, the centrality of social network analysis (Eq. 5) is used to measure how early adopters is close to the center of the network to decide on how important the early adopters is. The result is shown in Table 2.

Table 2. Social Influence of Roams

Early Adopter	Social Influence	Social Position
Netbook-1	-0.041	4
Netbook-3	-0.152	1
Netbook-6	-0.083	3
Netbook-7	-0.111	2

The results indicate that the key early adopters tried to create the blue ocean of useful daily life, in which he cannot only enjoy his leisure life with his virtual friends and watch online movie or TV, but also, if needed, can work while traveling.

5 Conclusion

Findings of their study show that our researching model can not only explore the innovative use of new products by the early adopters, but also break through the transformation of clusters of mostly bought product to consuming tribes. After connecting consuming tribes, as Granovetter's story illustrates, friends who were not so close served as the weak tie (bridge), helped him connect with other clusters (tribes) and gave him a different piece of information. Therefore, the discovered innovative ways of using a new product will bring a majority to accept new products in the future.

References

1. Agrawal, R., Imielinski, T., Swami, A.N.: Mining association rules between sets of items in large databases. In: SIGMOD 1993, Proceedings of the 1993 ACM SIGMOD International Conference on Management of Data, vol. 2, pp. 207–216. ACM, New York (1993)
2. Schindler, R.M.: Internet forums as influential sources of consumer information. Journal of Interactive Marketing 15(3), 31–40 (2001)
3. Cova, B.: Community and consumption: Towards a definition of the "linking value" of product or services. European Journal of Marketing 31(3/4), 297–316 (1997)
4. Granovetter, M.: The strength of weak ties: a network theory revisited. Sociological Theory 1, 201–233 (1983)
5. Hanneman, R.A., Riddle, M.: Introduction to Social Network Methods. University of California, Riverside (2005), `http://www.faculty.ucr.edu/~hanneman/`
6. Hirschman, E.C.: Innovativeness, novelty seeking and consumer creativity. The Journal of Consumer Research 7(3), 283–295 (1980)

7. Hong, C.-F.: Qualitative chance discovery: Extracting competitive advantages. Information Sciences 179(11), 1570–1583 (2009)
8. Ibarra, H.: Network centrality, power, and innovation involvement. Determinants of technical and administrative roles. Academy of Management Journal 36(3), 471–501 (1993)
9. Moore, G.A.: Crossing the chasm — Marketing and selling technology products to mainstream customers, 10th edn. Harper Collins, New York (1991)
10. Morrison, P.D., Roberts, J.H., von Hippel, E.: Determinants of user innovation and innovation sharing in a local market. Management Science 46(12), 1513–1527 (2000)
11. Ohsawa, Y., Nels, E.B., Yachida, M.: KeyGraph: Automatic indexing by co-occurrence graph based on building construction metaphor. In: Proceedings of 5th International Forum on Research and Technology Advances in Digital Libraries, pp. 12–18. IEEE Press, Santa Barbara (1998)
12. Ohsawa, Y.: Modeling the Process of Chance Discovery. In: Ohsawa, Y., McBurney, P. (eds.), pp. 2–15. Springer, New York (2003)
13. Price, L.L., Ridgway, N.M.: Use innovativeness, vicarious exploration and purchase exploration: Three facets of consumer varied behavior. In: Proceedings of the 48th Educator's Conference, pp. 56–60. American Marketing Association, Chicago (1982)
14. Robertson, T.S.: Innovative behavior and communication. Holt, Rinehart and Winston, Inc., New York (1971)
15. Rogers, E.M., Shoemaker, F.F.: Communication of innovations: A cross-cultural approach, 2nd edn. Free Press, New York (1971)
16. Rogers, E.M.: Diffusion of Innovations. Free Press, New York (2003)
17. Strauss, A.C., Corbin, J.M.: Basics of qualitative research: Techniques and procedures for developing grounded theory, 2nd edn. Sage Publications Inc., Thousand Oaks (1998)
18. Venkatesh, A., Vitalari, N.P.: Computing technology for the home: Product strategies for the next generation. Journal of Product Innovation Management 3, 171–186 (1986)
19. von Hippel, E.: Lead Users: A Source of Novel Product Concepts. Management Science 32(7), 791–805 (1986)

Data-Driven Innovation Technologies for Smarter Business from Innovators' Market Game to *iChance* Creativity Support System

Hao Wang and Yukio Ohsawa

School of Engineering, Department of Systems Innovation
The University of Tokyo, Bunkyo-ku, 113-8563 Tokyo, Japan
{wanghao0423,y.ohsawa}@gmail.com

Abstract. Innovation is often perceived as an unmanageable phenomenon, so most enterprises are struggling with a manageable and systematic approach for generating game-changing product ideas with the aggressive revenue growth goals. In recent years, customer knowledge has been gradually recognized as a vital source for enterprise success. Although many existing technologies on innovation, such as open innovation, innovation games, and gamestorming, are available for enterprise innovation, fewer valuable ideas are obtained finally. This chapter firstly reviews a novel data-driven innovation technology, Innovators' Market Game (IMG), which has been developed as a new form of innovation game based on chance discovery theory. Secondly, 4W-IMG is proposed as an extension(4W stands for Who, What, When and Where). For more acceptable ideas generation, a breakthrough tri-circle innovation model is built. Based on this model, a new systematic approach named Market Innovation Storming (MIS) is created. In addition, design teams in the enterprises often work in a distributed fashion across different time zone and location. Therefore, *iChance*, a Web-based collaborative role-playing creativity support system, is designed and developed, where 4W-IMG and MIS are separately implemented in *iChance* regular version and business version. Case studies indicate the effectiveness of systematic approach and *iChance* system.

Keywords: Creativity support system, Innovators' Market Game(IMG), *iChance*, Market Innovation Storming(MIS).

1 Introduction

In the past few years, business innovation (such as product or service innovation) has changed from traditional innovation model which only depends on internal source of a company, to open innovation model where end users of a company can be considered as a potentially beneficial source for their innovations. Open innovation can help firms to capture creative external ideas and valuable knowledge to create value beyond resources and capabilities of the firms [1]. In today's ultra-competitive market, Chesbrough [2] claims that organizations must take an open

Y. Ohsawa & A. Abe (Eds.): Advances in Chance Discovery, SCI 423, pp. 107–125.
springerlink.com

innovation approach by combining internal and external ideas, thereby gaining a competitive advantage. Dell, for example, is well known for its open innovation strategy where it supplies end users an online community, Dell IdeaStorm, to share their ideas and collaborate with Dell to create or modify new products and services [3]. Procter&Gamble (P&G) is another great example of open innovation and it has created a new innovation model of *Connect and Develop*, which achieves more than 35% of highly profitable innovations and billions of dollars in revenue [4]. In fact, P&G open innovation is based on collaborations with its customers and experts outside its own R&D department, achieving highly profitable innovations [5,6].

User-led innovation is a useful and effective specified open innovation [7,8]. Case studies of user-led innovations show how the users can be involved in the design and collaborative with firms for innovation [9]. However, for high technology products, users give little help due to lack of real world experience. von Hippel [10] proposes the new concept of "lead users" as a useful source of knowledge for understanding user needs and generating new product ideas based on those needs, which can accelerate product modifications and predict future demands. So, innovative complex products or services are developed by firms which benefit from applying "lead users" to create and capture new value.

Creative ideas for product concept development are very important in the early stage of new product development (NPD). NPD is a knowledge intensive process where the generation of new ideas and concepts requires detailed knowledge of both designers and customers [14]. As the matter of fact, breeding new and diverse ideas for new markets has always been a challenge as it often requires a nonlinear thinking approach [15]. Although innovation is usually seen as an unmanageable phenomenon, Miller et al. [16] have found innovation becomes manageable when managers get rid of universal prescription and recognize that different rules and practices apply in different contexts. In other words, innovation takes place in "innovation games", focusing on market creation and maintenance, and innovators support.

In current years, innovation games are becoming more and more popular in companies for business innovation. Innovation games create an environment for creative thinking and innovation and help companies collaborate with customers to break down barriers, communicate better, and generate new ideas, insights, and strategies [18]. Hohmann [17] has established Innovation Games Company and created a series of innovation games which help firms to get actionable insights into customer needs and preferences to improve performance through collaborative play.

It's not hard to see that previous innovation games are all based on a fixed and general scenario graph such as drawing a tree, a boat or a table on a white board, to solve any firm problems on how to understand customers and create new product ideas. However, it's much more scientific and effective way for enterprise innovation that scenario graph should be generated by analyzing real data from or about a company itself. And on the basis of scenario graph and a designated game rule, the company makes innovation in a game atmosphere through interaction with their customers. Therefore, Ohsawa [20,21]created a novel innovation game called Innovators' Market Game (IMG) to enhance human value cognition to scenario graph generated by Keygraph, therefore speeding up chance discovery process. Chance discovery, proposed by Ohsawa in 2000 [19], is a relatively new research field as an extension of data mining and knowledge discovery, which is a human-computer interaction process to detect rare but important chances for decision making. A chance means to understand an unnoticed event or situation which might be uncertain but significant for a decision. KeyGraph as a vital tool of chance discovery can generate scenario map to aid human value cognition in the process of chance discovery.

For companies, design teams often work in a distributed fashion, across boundaries of time and place (across locations, firms and countries), and their customers are in different geographical positions as well. Thus it's imperative to shift gaze from table game to Web game. Ohsawa et al. [21] proposed Web IMG, and called it Web Innovators Marketplace (IM), as it does not provide gaming air in this Web version. Therefore, a Web system with a systematic approach in a game environment for innovation is still eager to be established.

In this chapter, innovators' market game(IMG) is reviewed and a case study of IMG is presented in section 2. Section 3 introduces 4W-IMG as an extension of IMG. For more valuable ideas creation, Market Innovation Storming(MIS) with a new innovation model is presented in section 4 and conclusion is in section 5.

2 A Case Study of Innovators' Market Game (IMG)

2.1 Introduction to IMG

Innovators' Market Game has been developed as a designed workplace for conversation to create innovative ideas [19,20].

Materials Preparation. When setting up a game of IMG, some necessary materials are offered to players as below:

- **Basic Card**. Each basic card records the basic information of a product.
- **Virtual Money**. Players purchase ideas or basic cards by virtual money.
- **Demand Post-it Note**. Customers record their demands in the demand post-it note.
- **Idea Post-it Note**. Inventors write down a creative idea in each idea post-it note. Each innovator has specified color of post-it note to record.
- **Game Board**. In scenario graph as game board, a black node corresponds to a basic card and red nodes imply innovative ideas created by inventors.
- **Role Card**. Each role card represents a customer base, such as student, doctor, engineer, or housewife, etc.

The Roles of Players There are three roles, of which the core role is the inventor.

- **Facilitator.** One participant is selected to be the facilitator who assists all players to carry out IMG smoothly. Facilitator needs to hand out the same account of virtual money to each player before IMG starts and sell basic cards to inventors during the game. Facilitator also needs to control the playtime and encourage inventors to create more ideas all the time.
- **Inventor.** Choose 3 or 4 participants to be inventors who represent product designers from different companies. They have to purchase basic cards from the facilitator before they prepare to create new ideas. There are two ways to purchase. One way is the facilitator randomly distribute cards to customers; the other way is inventors buy cards as they appoint, but the purchase price of each card will be doubled. Once inventors receive basic cards, they may combine these cards to create new ideas according to scenario graph on the game board. Inventors must write down each new idea on an idea post-it note, and then attach it onto the game board. They also need to loudly announce their innovative ideas to all players.
- **Consumer.** Customers can choose their own roles such as student, teacher, or housewife with the role cards provided for the game. Customers may purchase their creative ideas from inventors or propose new demands on their demand post-it notes. They also need to loudly announce their demands to all players and attach post-it notes onto the game board.

The Rule. After IMG starts, inventors buy basic cards from the facilitator, and then try to create new product ideas by combining basic cards, according to each red node's position and relationship with other black nodes on the game board. Meanwhile, customers are encouraged to put forward as many new demands as possible. Once customers have new demands, they should write them down on the demand post-it note. Inventors and Customers both must loudly announce their ideas and demands to all the players and attach post-it notes on the appropriate position of game board. IMG usually lasts for 1-2 hours. Consumers and inventors communicate each other and co-create innovative ideas so that potential business ideas emerge.

Evaluation. When the game is over, the inventor who earns the most money will be the winner among all inventors. Similarly, the customer who obtains more votes from other players wins the game among consumers.

2.2 A Case Study of IMG

An experiment of IMG was conducted in a special section of an international symposium with the aim of generating new product ideas for Daiso Shop, a convenient chain store in Japan [23]. Teamwork was essential for co-creation and had stonger power for team innovation. Thus there were three inventor groups, and each group had two inventors, see Fig. 1.

Fig. 1. Three roles of players

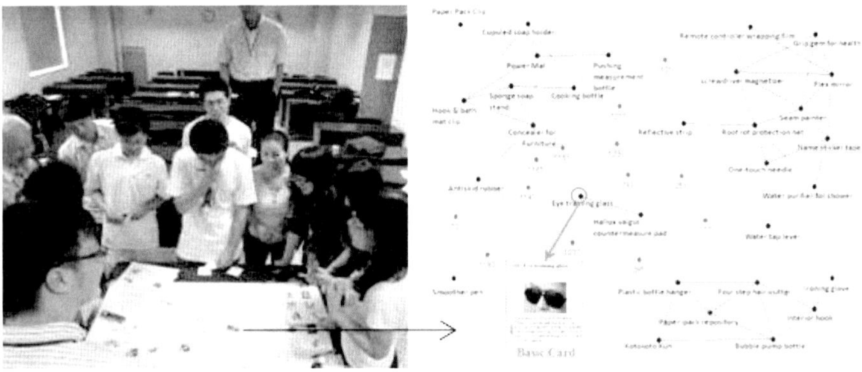

Fig. 2. A scene of IMG and a view of game board

Thirty basic cards were made for thirty different products. The right graph of Fig. 2 shows scenario graph was generated by Keygraph tool to visualize the information of these products. About twenty players joined in the IMG, and they were VIP professors and student participants of the conference.

After one hour of playtime, six creative ideas were obtained from inventors and six demands were proposed by consumers. Fig. 2 shows a scene of IMG: inventors are standing in the center, and customers are around them.

3 4W-IMG

4W-IMG, a customer-centric innovation game is designed and developed based on original IMG. The details of 4W-IMG approach is illustrated as below.

3.1 Materials Preparation

In 4W-IMG,the product certificate as a new material is provided to inventors and customer role card is removed. Customers purchase creative ideas from inventors and inventors issues product certificates to customers for their valuable demands or complaints. Moreover, customers can propose any demands and represent any customer roles as they wish.

- **Basic Card.** A basic card records a piece of product information.
- **Virtual Money.** Customers purchase new product ideas from inventors by virtual money.
- **Demand Post-it Note.** Customers record each demand in a demand post-it note.
- **Idea Post-it Note.**Inventors record each creative idea on a idea post-it note. Each inventor has specified color of idea post-it note.
- **Game Board.** The same as defined in IMG.
- **Product Certificate.** Inventors award product certificate to customers for their valuable requirements.

3.2 The Roles

4W-IMG has the same three roles as IMG: Customer(Consumer), Designer (Inventor) and Facilitator, where the core role is the Customer in 4W-IMG instead.

3.3 The Rule

4W-IMG is conducted in the following steps:

Step 1: According to basic cards and scenario graph, customers ought to propose new demands and write them down on demand post-it notes. Each post-it note should only record one demand, and is required to be written in accordance with the format of 4W-demand consisting of What, Who, When and Where, see Fig. 3. "What" means requirement or complaint customer proposes. "Who" refers to a class of customer group (such as student, teacher, etc) a customer wants to act as to while proposing a demand. "When" and "Where" refer to time and location of demand application. Finally, customers should loudly announce their requirements to inventors and then attach demand post-it notes onto the suitable positions of the game board. In addition, customers can use virtual money to purchase innovative ideas from inventors.

Step 2: Inventors are advised to create new ideas which correspond to red nodes shown in scenario graph by combining black nodes linking these red nodes. They need to pay much more attention to customers' 4W demands, and try to discover what customers really want or don't like about their offerings. As shown in Fig. 4, designers need to record creative ideas by "How" and "Why" called H-W idea. That means designers need to not only explain their ideas as detailed as possible but also explain the reasons for their ideas. When inventors adopt a requirement or complaint from a customer and create a new idea, they should award a product certificate to that customer for the contribution.

Number & Name: (Customer)
What: *demands or complaints*
Who: *customer base*
When: *time*
Where: *place*

Fig. 3. 4W-demand/complaint of a customer

Number & Name: (Inventor)
How: *design ideas or concepts*
Why: *reason for design*

Fig. 4. H-W idea for designer

3.4 Evaluation

The evaluation method of IMS is designed with the purpose of high-quality ideas generation.

Customer. Calculate the total number of product certificates of each customer by Eq. 1 and then sort them.

$$C\_Vote(i) = \Sigma Vi \tag{1}$$

where Vi means the number of product certificates the ith customer obtains.

Designer. The performance of each designer is assessed by the total number of virtual money they have earned. However, for driving designers to create more high-quality ideas, the final score of each designer should be calculated by Eq. 2

$$D\_Score(i) = \frac{Ti}{Ni} \tag{2}$$

where Ti indicates the total number of virtual money the ith designer earns; Ni is the number of ideas the ith designer creates.

At last, two winners are chosen: one inventor and one customer.

3.5 Web 4W-IMG: *iChance* Regular Version

For companies, design teams often work in a distributed fashion, across boundaries of time and place, and their customers are in different geographical positions as well. Thus it's imperative to shift gaze from table game to Web game.

iChance, a Web-based collaborative role-playing creativity support system (CSS), has been designed and developed [25,26,27]. *iChance* is a dynamic interaction platform for product designers to make collaborative market innovation with their customers on the Web. The whole operating procedure of *iChance* is in accordance with 4W-IMG approach, which can improve users' insights to scenario graph, capture clear voice of customers (VOCs) and speed up the creation process of creative ideas on the Web platform.

System Interface. Fig. 5 shows the interface of *iChance*, including four main functional modules: (1) scenario-based innovation module, (2) requirement module, (3) communication module and (4) toolbar module. In Module (1), a scenario graph is provided to users for new value cognition and creation. Designers can create new ideas put in appropriate positions of the graph. Module (2) is for customers to propose their new demands. Customers and designers communicate each other in Module (3). Some useful tools for user action are in Module (4).

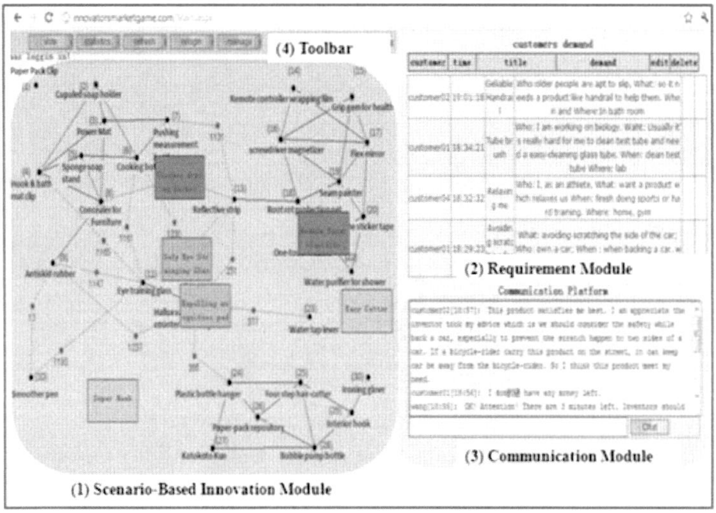

Fig. 5. The Interface of *iChance* (regular version)

Roles and their actions on Web 4W-IMG. In Web 4W-IMG system, different roles are granted different operation authorization when users log in. The following content will introduce the responsibility of each role and how to operate and interact between these roles on the Web platform.

Facilitator. A facilitator should assist users to learn 4W-IMG approach and system operation. As IMG goes on, the facilitator encourages customers to propose as many demands as possible, thereby inspiring designers to create innovative ideas. In fact, facilitator is an extremely important role who impacts the quality of final result.

Customer. According to scenario graph, customers should release their demands or complaints in the requirement module, shown in Fig. 6. Each customer is required to record a requirement at a time and must follow the specified format of 4W demand.

Designer. The primary responsibility of designers is to create breakthrough products or services through making value cognition to scenario graph and interaction with customers. They are advised to create new ideas which correspond to red nodes shown in scenario graph by combining black nodes linking these red nodes. Meanwhile, they need to pay much more attention to customers 4W demands, see Fig. 7, and try to discover what customers really want or don't like about their offerings (ideas).

Interaction between customers and designers. There are some interactive behaviors between customers and designers. They can exchange views in communication module. In addition, customers may score any idea created by designers if they think the idea is good, see Fig. 8. And designers can vote for customers, whose demands are significant for their idea generation, see Fig. 9.

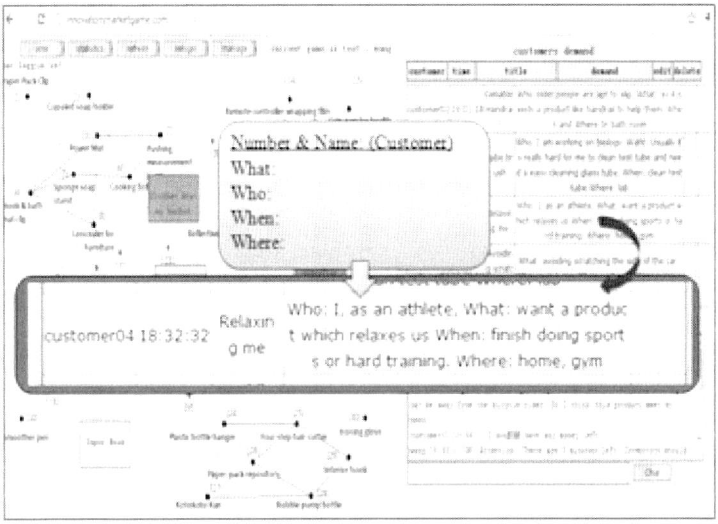

Fig. 6. 4W demand proposed by a customer

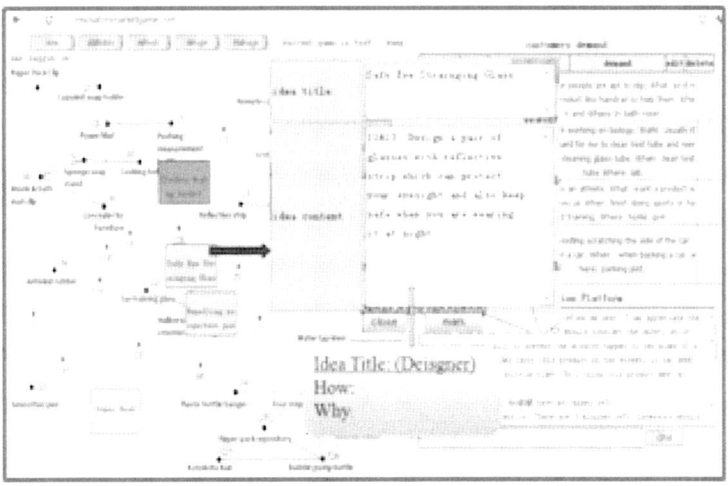

Fig. 7. H-W idea from a designer

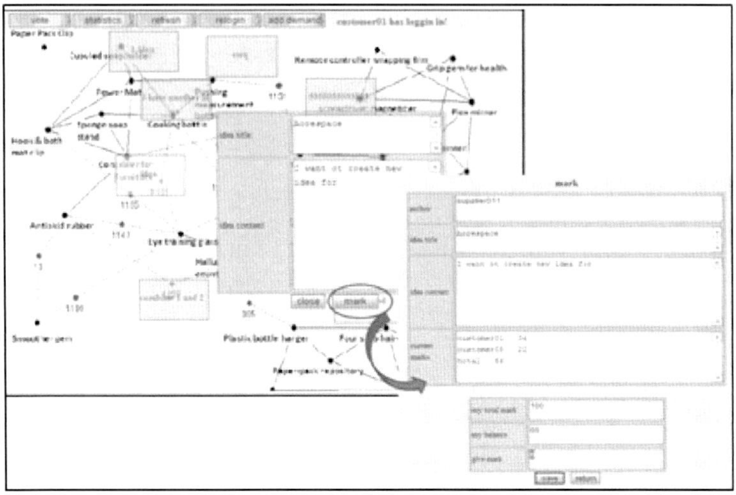

Fig. 8. The customer marking an innovative idea created by a designer

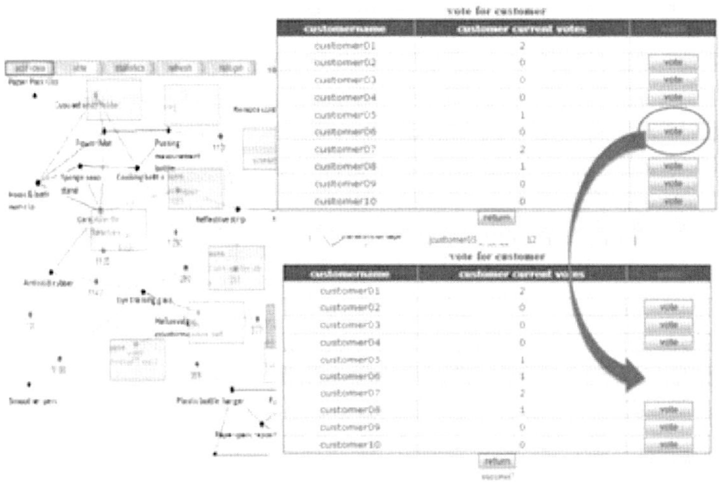

Fig. 9. A designer voting for a customer for the valuable demand

4 Market Innovation Storming(MIS): An Evolution of Innovation Technology

Many relevant techniques have been proposed to foster creativity of individuals or groups, such as brainstorming, mind mapping, morphological analysis, patent map-ping, etc. These technologies for innovation, including aforementioned new methods: IMG and 4W-IMG, separately employ different scenario graph, different interactive rule, different role-centric playing or idea source, but a common

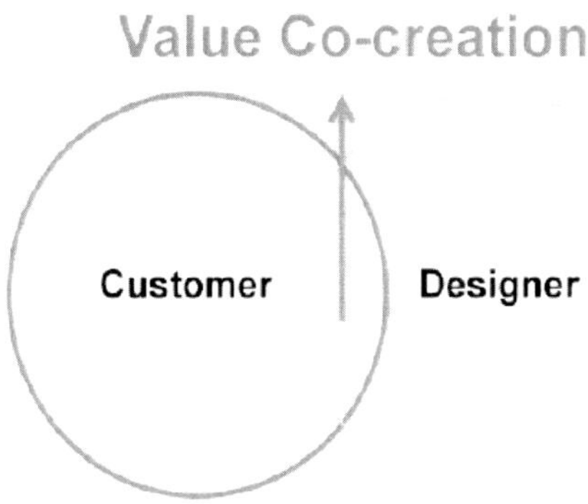

Fig. 10. Value co-creation model in current innovation technologies

innovation pattern can be discovered that new product ideas emerge through co-operation and communication between enterprises and their customers. In other words, new value is co-created by interaction between designers and customers, see Fig. 10.

4.1 Tri-circle Innovation Model

Research has defined some stages as the "fuzzy-front-end" of the product innovation process before a new product is decided to develop, even though firms have their own product innovation processes [28]. Several studies have shown that product innovation activities conducted at the fuzzy-front-end of product innovation process identify successful and unsuccessful new products based on the most important factors [11,12,13]. However, it.s widely recognized for firms that fewer ideas are eventually accepted to invest in, when technical experts further evaluate these ideas generated by these innovation technologies.

Therefore, a problem still remains on how to build an effective breakthrough method for more acceptable ideas generation.

A tri-circle innovation model is developed. As shown in Fig. 11, the union of the three circles (customer, designer and technical expert) represents the most acceptable high-quality ideas generated with the consideration of customer's demands or complaints, designer's innovation and technical expert's assessment.

4.2 Market Innovation Storming (MIS)

Based on the tri-circle innovation model, a new method of Market Innovation Storming (MIS) is designed for

- generating more acceptable ideas;
- obtaining clear voice of customers and creative ideas of designers;
- accelerating the innovation process.

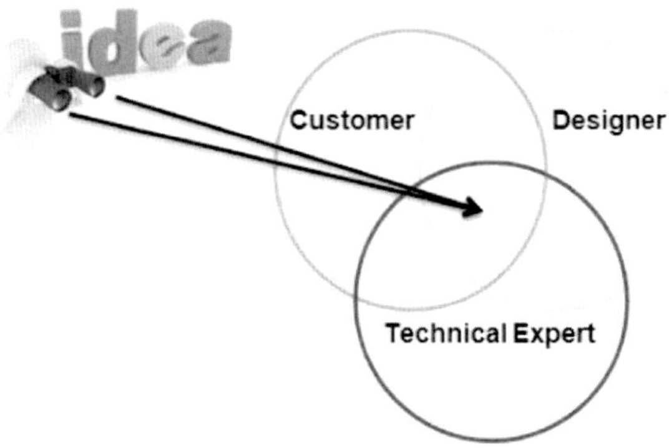

Fig. 11. Tri-circle innovation model

The Roles Similar to 4W-IMG, MIS is also a customer-centric innovation process where technical expert is added as a vital role.

- **Customer.** Propose potential demands or complaints.
- **Designer.** Create innovative ideas.
- **Technical Expert.** Verify ideas from designers, such as the realizability, cost, development cycle, etc, and point out valuable information to customers and designers.
- **Facilitator.** Manage and control entire innovation process.

Methodology. Fig. 12 shows the interactive process of MIS among customers, designers and technical experts.

Step 1: Fully understand the objective, collect relevant data and make scenario graph at last. Facilitator prepares to set up MIS.

Step 2: Based scenario graph, customers propose new demands or complaints which should be recorded with 4W-Demand format introduced in 4W-IMG. Demands or complaints are grouped into internal and external ones. If proposed demands or complaints are related to scenario graph, they are regarded as internal demands or complaints and vice versa.

Step 3: Technical experts recommend useful complaints or potential demands to designers. In Fig.14, labeled complaints and demands means these requirements or complaints are strongly recommended to designers.

Step 4: Based on scenario graph and customer demands or complaints, designers create new idea (solution or strategy) with H-W idea format. Similar to demands or complaints, ideas also have two categories: internal and external. If an idea is generated due to scenario graph, it is an internal idea, and vise versa. If possible, designers make the design sketching for clearly expressing their ideas.

Step 5: Technical experts validate each idea from designers. If an idea is considered valuable and practical, technical experts will further give technical information about this idea, such as applied technology, development cycle, cost, etc. Validated ideas will be suggested to customers.

Step 6: Customers and designers communicate and interact with each other about their demands/complaints or innovative ideas. Customers will score their favorite ideas, and designers will mark useful complaints/demands for their new ideas generation.

Step 7: Repeat Step 2 to 5.

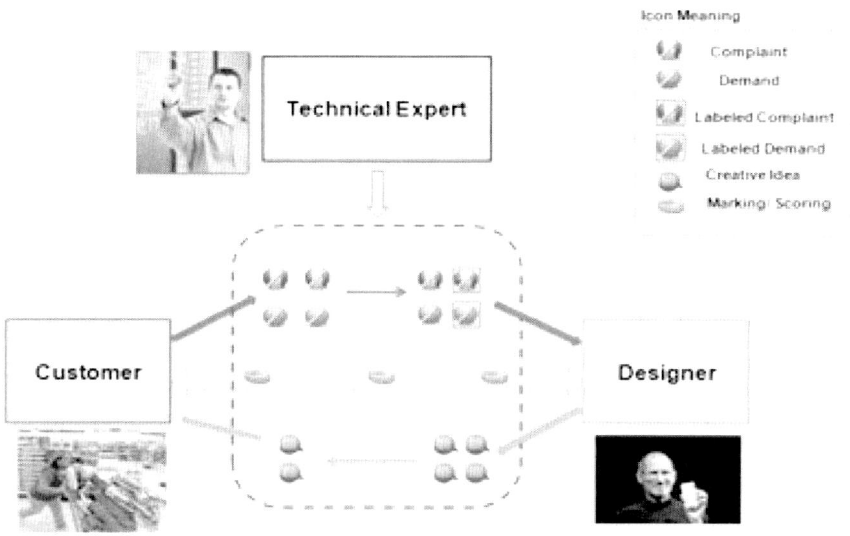

Fig. 12. MIS interactive process among customers, technical experts and designers

4.3 Web MIS: *iChance* Business Version

The approach of MIS has been implemented in *iChance* business version. Compared with regular version, *iChance* business version provides more powerful functions for better user interaction and experience.

The interface. The layout of the interface in these two versions is very similar. As shown in Fig. 13, there are four modules as below:

1. **Scenario graph module.** It provides users to make insights and cognition. Users can obtain useful basic information, publish their views and evaluate each other's comments.

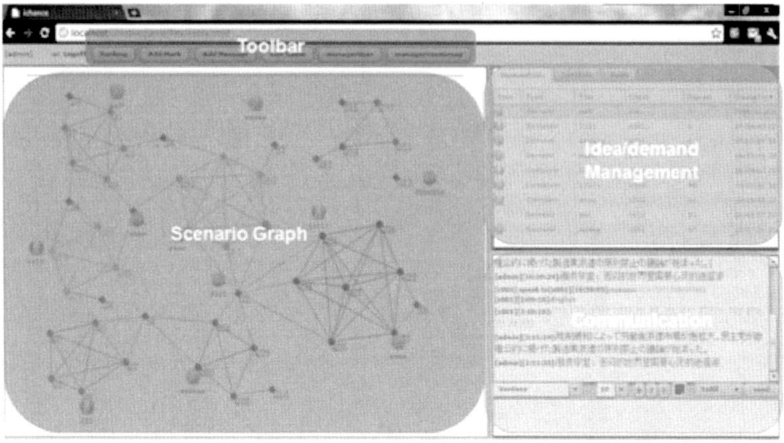

Fig. 13. The interface of *iChance* business version

2. **Knowledge management module.**Ideas and demands as important knowledge are managed in this module. Besides, a ranking function is provided to users to check.
3. **Communication module.** Users can communicate each other by using different font, color or size.
4. **Toolbar.** Some function buttons are in this module for users to operate.

The operation and interaction of roles. There are four roles in in *iChance* business version: facilitator, customer, designer and technical expert.

Facilitator. The facilitator should configure necessary information in *iChance* before they start a MIS for an innovation activity. First, they should register basic information for each user, set username and password for them to log in system and group them into different role. Second, they need to fill up necessary information, such as project name, project content, and upload scenario graph into the system.

Customer. For a demand pop-up box shown in Fig. 14, customers firstly should determine a demand or complaint they want to propose and fill up 4W demands or complaints. They should make sure if proposed demands or complaints are shown in scenario graph at the end. That is, customers need to identify if the demand or complaint is internal or external. Customer needs to drag the demands and complaints to a proper position in scenario graph. All demands or complaints are recorded in the demand/complaint list of knowledge management module.

Designer. The designer creates a new idea by how and why, and gives a title of the idea. They also need to indicate the idea is either internal or external. Most importantly, designers should upload design sketching for better

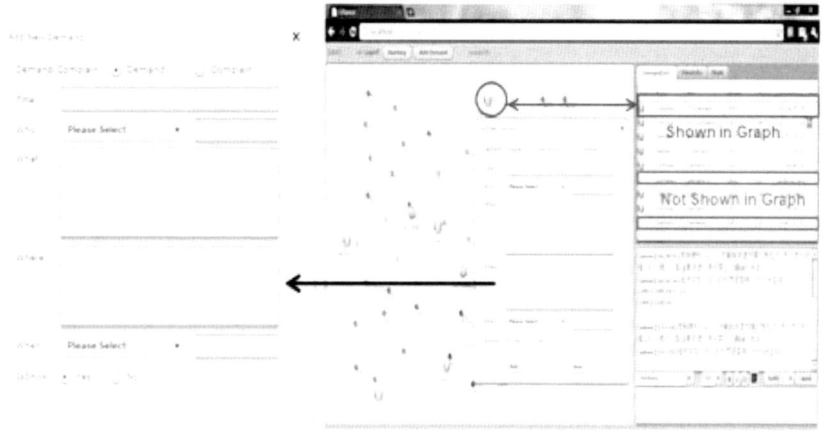

Fig. 14. A customer proposing a demand or complaint in *iChance*

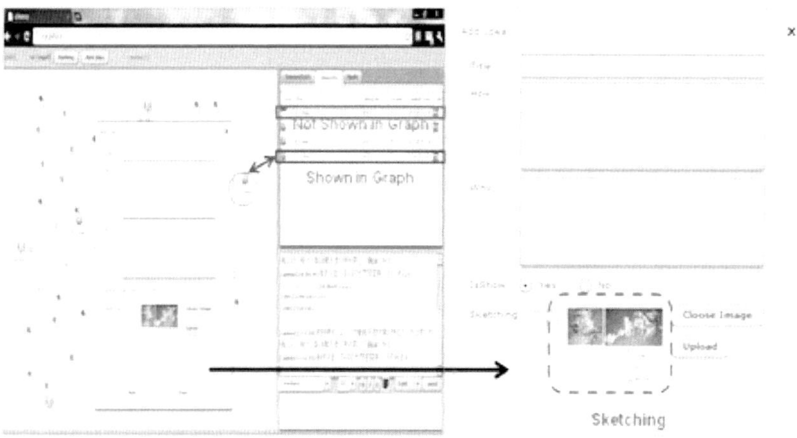

Fig. 15. A designer creating an idea in *iChance*

describing their ideas, see Fig. 15. All ideas are shown in the idea list of knowledge management module.

Technical expert. Technical expert takes charge of checking the realizability and effectiveness of creative ideas. Once an idea is proved, technical expert will further provide necessary technical information, such as development circle, costs, and relevant technologies, see Fig. 16. In addition, technical experts also can recommend useful complaints or potential demands to designers.

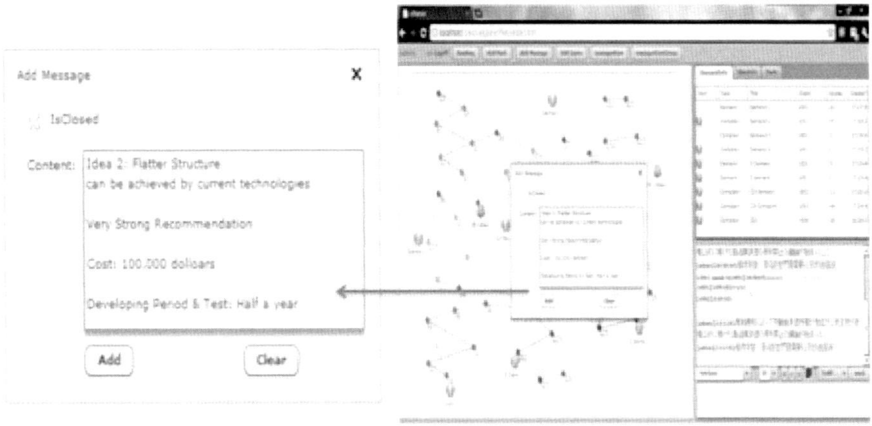

Fig. 16. Technical expert verifying creative ideas

The interaction between customers and designers. As shown in Fig. 17, there are two ways for designers to mark customer demands or complaints. One is designers click the icon of a demand or complaint in the scenario graph module, and an information box pops up. Designers can read the details of the demand or complaint and then mark it. The other is click the item in demand information list of knowledge management module, the same pop-up information box shows up for designers to read and mark. It must be pointed out that the first way is only for marking internal demands or complaints. Fig. 18 shows customers have the similar operation to score the innovative ideas from designers.

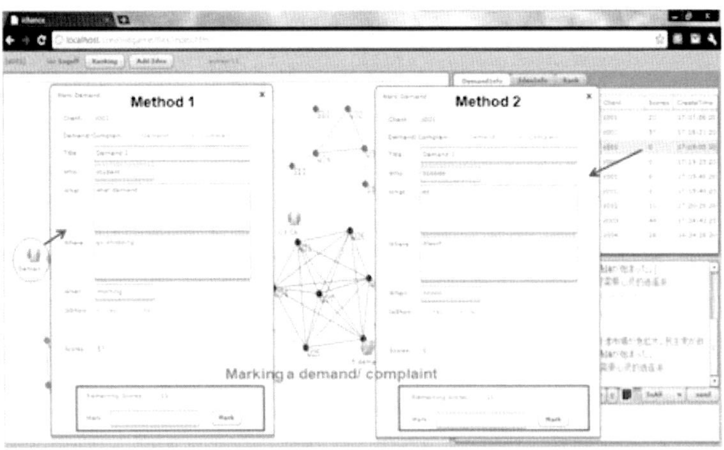

Fig. 17. The designer marking a demand or complaint

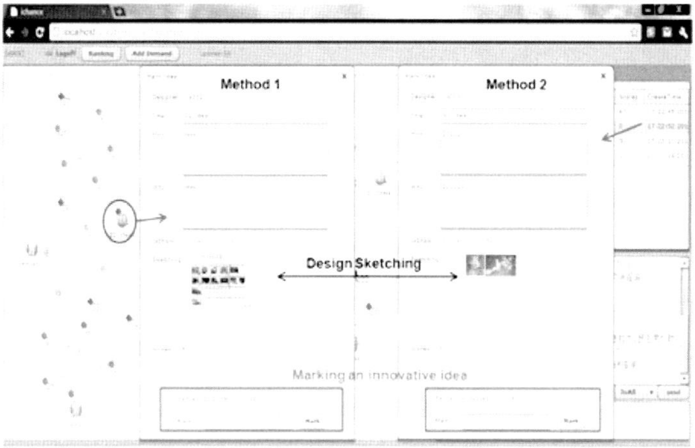

Fig. 18. The customer scoring an innovative idea

4.4 A Case Study on *iChance* Business Version

A successful case study of *iChance* business version has been carried out in a real company, one of top 500 enterprises in the world. The company wants to effectively capture more valuable ideas to support their next-generation interactive media system for further development through game-changing innovation.

For the purpose of comparison, game storming (GT), *iChance* regular version (*iChance*-R) and business version (*iChance*-B) are separately employed to generate creative ideas. Game storming is an innovation tool of the company. Fig. 19 indicates that *iChance*-B has the highest idea accepted rate of nearly 90%, even though the total number of generated ideas is less than the other two methods'. *iChance*-R achieved the most number of ideas with the idea accepted rate of a little bit more than 50%. The company's traditional innovation tool GS has less than 40% idea accepted rate.

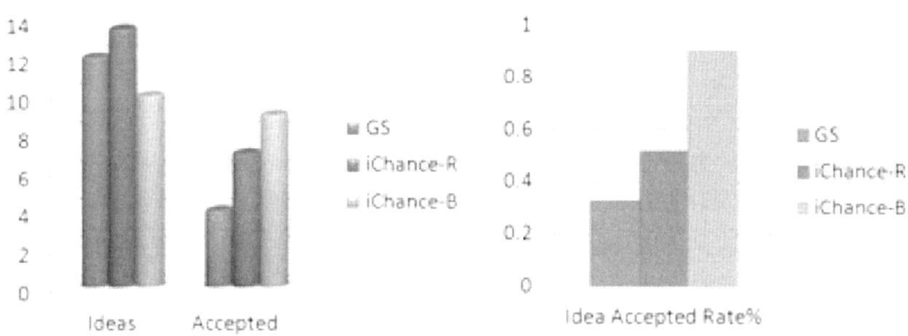

Fig. 19. The comparison of GT, iChance-R and iChance-B

5 Conclusion

Many studies on innovation have paid more attention to customer knowledge, but fewer valuable ideas are decided by firms to invest in. Most innovation methods follow a general pattern: set a topic or draw a picture on a white board, and then create ideas through group meeting or workshop. However, data-driven innovation technologies are regarded as the best practice for firms. According to the practical problems enterprises encounter, select relevant real world data and then generate scenario graph by analyzing the data. At last, creative ideas are emerging when users make insights to scenario in a game-like environment.

This chapter presents 4W-IMG approach based on original IMG. 4W-IMG is a customer-centered innovation process where 4W demands as clear voice of customer are obtained from customers and innovative H-W ideas are created by designers through their interaction. For more valuable ideas generation, market innovation storming (MIS) is developed with a breakthrough innovation model where technical expert as an important role is put into the interactive process. Finally, a creativity support system named *iChance* is built based on these two systematic approaches of 4W-IMG and MIS.

iChance has increasingly attracted more and more attention from academia and business. Some companies have expressed potential interests in *iChance* for their future innovation. Therefore, more case studies will be done in these companies.

References

1. Chesbrough, H.W.: Open Innovation: The New Imperative for Creating and Profiting From Technology. Harvard Business School, Boston (2006)
2. Chesbrough, H.W.: The era of open innovation. MIT Sloan Management Review 44(3) (2003)
3. Di Gangi, P.M., Wasko, M.: Steal my idea! Organizational adoption of user innovations from a user innovation community: A case study of Dell IdeaStorm. Decision Support Systems 48, 303–312 (2009)
4. Huston, L., Sakkab, N.: Connect and develop. Harvard Business Review 84, 58–66 (2006)
5. Chesbrough, H.: Open business models. Harvard Business School Press, Boston (2007)
6. Hau, Y.S., Kim, Y.G.: Why would online gamers share their innovation-conducive knowledge in the online game user community? Integrating individual motivations and social capital perspectives. Computers in Human Behavior 27, 956–970 (2011)
7. Freeman, C.: Networks of innovators: a synthesis of research issues. Research Policy 20, 499–514 (1991)
8. von Hippel, E.: The dominant role of users in the scientific instrument innovation process. Research Policy 5, 212–239 (1976)
9. Ornetzeder, M., Rohracher, H.: User-led innovations and participation processes: lessons from sustainable energy technologies. Energy Policy 34, 138–150 (2006)
10. von Hippel, E.: Lead users: a source of novel product concepts. Management Science 32, 791–805 (1986)

11. Cooper, R.G., de Brentani, U.: New industrial financial services: what distinguishes the winners. Journal of Product Innovation Management 7, 75–90 (1991)
12. De Brentani, U.: Success and failure in new industrial services. Journal of Product Innovation Management 6, 239–258 (1989)
13. Ozer, M.: The roles of product lead-users and product experts in new product evaluation. Research Policy 38, 1340–1349 (2009)
14. Sofinati, T.D., Suryadi, K., Govindaraju, R., Prihartono, B.: New product development based on customer knowledge management. In: APIEM 2009, Kirakyushu, pp. 1268–1279 (2009)
15. Kim, S.K., Ishii, K., Beiter, A.K.: Scenario graph: discovering new businesses opportunities and failure modes. In: Proceedings of ASME International Design Engineering Technical Conferences, Las Vegas, Nevada, pp. 603–611 (2007)
16. Miller, R., Olleros, X., Molini, L.: Innovation Games: A New Approach to the Competitive Challenge. Long Range Planning 41, 378–394 (2008)
17. Hohmann, L.: Innovation Games: Creating Breakthrough Products Through Collaborative Play. Addison-Wesley (2006)
18. Gray, D., Brown, S., Macanufo, J.: Gamestorming: A Playbook for Innovators, Rulebreakers, and Changemakers. O'Reilly Media (2010)
19. Ohsawa, Y., McBurney, P. (eds.): Chance Discovery. Springer, Heidelberg (2003)
20. Ohsawa, Y.: Innovation Game as a Tool of Chance Discovery. In: Sakai, H., Chakraborty, M.K., Hassanien, A.E., Ślęzak, D., Zhu, W. (eds.) RSFDGrC 2009. LNCS, vol. 5908, pp. 59–66. Springer, Heidelberg (2009)
21. Ohsawa, Y., Okamoto, K., Takahashi, Y., Nishihara, Y.: Innovators Marketplace as Game on the Table versus Board on the Web. In: Proceeding of IEEE International Conference on Data Mining Workshops (ICDMW), Sydney, Australia, pp. 816–821 (2010)
22. Ohsawa, Y.: Data crystallization: chance discovery extended for dealing with unobservable events. New Mathematics and Natural Science 1(3), 373–392 (2005)
23. Wang, H., Ohsawa, Y., Nishihara, Y.: A system method to elicit innovative knowledge based on chance discovery for innovative product design. International Journal of Knowledge and Systems 2(3), 1–13 (2011)
24. Wang, H., Ohsawa, Y., Nishihara, Y.: Innovation Support System for Creative Product Design Based on Chance Discovery. Expert Systems with Applications 39, 4890–4897 (2012)
25. Wang, H., Ohsawa, Y.: iChance: A Web-Based Innovation Support System for Business Intelligence. International Journal of Organizational and Collective Intelligence 2(4), 48–61 (2011)
26. Wang, H., Ohsawa, Y.: iChance: Towards New-Generation Collaborative Creativity Support System for Advanced Market Innovation. In: The 6th International Conference on Knowledge, Information and Creativity Support Systems (KICSS), Beijing, China, pp. 138–143 (2011)
27. Wang, H., Ohsawa, Y.: Web-based Innovation Supporting System for Creative Ideas Emerging. In: The 6th International Workshop on Chance Discovery (IWCD6) in the 22nd International Joint Conference on Artificial Intelligence (IJCAI), Barcelona, Spain, pp. 68–76 (2011)
28. Rosenau, M.D., Griffin, A., Castellion, G., Anschuetz, N.F.: The PDMA Handbook of New Product Development. John Wiley & Sons, New York (1996)

Paired Evaluators Method to Track Concept Drift: An Application in Finance

Masabumi Furuhata[1], Takanobu Mizuta[2], and Jihei So[2]

[1] Computer Science Department, University of Southern California,
941 W37th Pl., Los Angeles, CA, 90089-0782, USA
k439bk439b@gmail.com

[2] SPARX Asset Management Co. Ltd., Japan,
Gate City Ohsaki, East Tower 16F 1-11-2 Ohsaki, Shinagawa-ku, Tokyo 141-0032, Japan
mizutata@gmail.com, Jihei.So@sparxgroup.com

Abstract. We consider the problem of forecasting under the environments of sudden unexpected changes. The objective of the forecasting is to detect several different types of changes and to be adaptive to these changes in the automated way. The main contribution of this paper is a development of a novel forecast method based on paired evaluators, the stable evaluator and the reactive evaluator, that are good at dealing with consecutive concept drifts. A potential application of such drifts is Finance. Our back-testing using financial data in US demonstrates that our forecasting method is effective and robust against several sudden changes in financial markets including the late-2000s recessions.

1 Introduction

The widely accepted models often encounter a sudden unexpected change of circumstances, in which there are instances when the original models become invalid. Typical reasons of such invalidations are classified into the following ways: (1) models do/can not include all important factors in models that have not been obvious; (2) sudden and great environmental changes lose effectiveness of models; (3) interactions of small changes have boosted as significant changes as a whole. With respect to these changes, it is difficult to predict before occurrences or even during the occurrences. In many cases, new important factors are observable in ex-post facto analysis; early discoveries of such factors tend to result in key success components in business. For decision makers, important decision items are the timing of discontinuance of their current models that are expected to loose their effects, and introduction of new models considering newly found factors that are expected to be effective.

Another approach to deal with these problems is to embed functions of detection or discovery of such changes and adoption to changes into a model which has gained attentions in data mining and machine learning. These changes over time in unforeseen ways are known as *concept drift* in these fields. A main difficulty to deal with concept drift is the greater number of observations does not simply lead to the increase of forecast accuracy unlike phenomena governed by laws of nature. Therefore, researches dealing with concept drift are extensive such as determination of window size [12,21], change detections [3,2,8,9], and adaptive ensembles [10]. Our research is most closely related

Y. Ohsawa & A. Abe (Eds.): Advances in Chance Discovery, SCI 423, pp. 127–141.
springerlink.com

to the determination of window size for a prediction. In this domain, there are two main streams to cope with concept drifts, dynamically changing the window size [12,21] or using two fixed window sizes [1,16]. In the former stream, as soon as they observe a new data, they investigate consistencies with the histories. Once they suspect an occurrence of concept drift, they adjust their window sizes. In the latter stream, they use paired classifiers to control two types of window sizes. A common point in the researches in both streams is that they adjust window size for classification problems.

In contrast, we propose an approach for forecasting problems that uses several different base forecasts having different window sizes simultaneously and the most appropriate window size is expected to be selected by an appropriate evaluator. Instead of adjustment of window size, evaluators in our approach are able to dynamically change the windows size by a selection of base forecasts. Therefore, a selection of an appropriate evaluator is a key issue in our approach. Similarly to the existing researches of the paired classifiers [1,16] to deal with concept drift for classification, we use two types of evaluators, a stable evaluator and a reactive evaluator, to deal with concept drift for prediction. While the stable evaluator is used as a default evaluator which is supposed to be appropriate for versatile situations, the reactive evaluator is sensitive to changes. If performances of the reactive evaluator exceeds ones of the stable evaluator, our method switches to use a base forecast selected by the reactive evaluator. With respect to decisions for switches, we use learning algorithm according to the histories of performances. A main characteristic of our proposing method is robustness against consecutive occurrences of concept drifts. We examine a back-testing using actual financial data in US in order to demonstrate how our proposing method performs compared with other existing methods.

This research is relevant to modern portfolio theory which is based on capital asset pricing model (CAPM) established by Sharpe [18], Lintner [13], and Mossin [14]. A main characteristic is an emphasis on a price discovery process rather than pricing itself. In the CAPM, a theoretically appropriate required rate of return of an asset is obtained according to a consideration of the expected return of the market, the expected return of a theoretical risk-free asset and non-diversifiable risk. Hence, the non-diversifiable risk is used as a single *factor* to compare the excess returns of a portfolio with the excess returns of the entire market that entails the set of optimal equities for a portfolio. More recently, Fama and French [6] propose two risk factors, *value* and *size*, and Carhart [4] proposes a factor, *momentum* that are widely accepted to reduce some exceptional cases of the CAPM:

- momentum: historical price increase for 12 months,
- value: book-to-market ratio,
- size: size of a firm (market capitalization).

Even though several factors have been proposed to predict future market movements, a persistent factor has not found yet. Hence, a key issue for investors based on factors is to select the best factor which suddenly and significantly changes over time.

Our financial strategy is *passive management* in which a factor selection does not involve any investor's intuition. Thus we aim to automate chance discovery process in the forecasting method. In addition, we do not rely on the external data such as macro economic statistics in order to be independent from noises contained in the external data. By contrast, in *active management*, fund managers make specific investments. This type of strategy is relevant to one stream of chance discovery that aims to integrate the chance discovery process by the interaction of human thoughts and automated data mining systems [15,17]. Unfortunately, our methodology cannot be applied to active management entirely, since human knows the back-testing data which we use in our experiment. However, our forecasting method can help active managers to provide information how the system detects market changes.

The rest of this paper is organized as follows. In Section 2, we show related work in the domain of concept drift. In Section 3, we detail our proposing forecasting method. Some key characteristics of our method are shown with some examples in Section 4. In Section 5, we examine a back-testing and compare the performances of our method and other representative methods. In Section 6, we describe conclusions of this paper.

2 Related Work

Concept drift is usually used for either prediction or classification. While a target for prediction is real number, a target for classification is an element in the set. Covering these two usages, Žliobaitė [20] provides an extensive literature review in the domain of concept drift. Our work is mostl relevant to learners with dynamic window size of forecast. Even though this paper focuses on prediction, we present some work of classification, since techniques between these two have some commonalities.

Žliobaitė [20] classifies typical concept drifts into four change types: (1) sudden drift, (2) gradual drift, (3) incremental drift, and (4) recurring contexts. While a quick response is significant for the sudden drift, a reaction to a change should be careful for the gradual drift. Regarding the incremental drift, it is important to detect small changes and take small adjustments at once. In contrast, with respect to the recurring contexts, it is crucial to find a similar occurrence in the history.

Regarding the first three types of concept drifts, an adjustment of forecast window size is a crucial issue. While recent events are taken into account as greater factors to react to sudden drifts with a smaller window size, frequencies of changes and speed of changes are extracted more accurately with a larger window size. For these issues, Lazarescu et al. [12] propose Competing Windows Algorithm (CWA) that adjusts the size of forecast windows according to characteristics of historical data. If a new observation is consistent over time, this algorithm uses larger windows size in order to increase forecast accuracy with an expectation of no concept drift occurrence. Otherwise, it uses a smaller window size to deal with concept drifts. In order not to focus too much on a particular window size, it uses three types of windows, small-medium-large, that are dynamically changing. For a forecasting, the most accurate forecast is used among the forecasts generated from the three windows.

FLORA is a representative learning system that deals with recurring contexts with its dynamically changing window size proposed by Widmer and Kubat [21]. While every step FLORA observes a new data, it searches relevant historical data and classifies the searched data into positive data, negative data and both type data. Based on these data,

FLORA generate forecast. According to results, FLORA updates its source of concepts: either addition of a new concept into the system or discard of the old concept.

Bach and Maloof [1] propose paired learners that consists of the stable learner that predicts based on all of its experience and the reactive learner that predicts based on its most recent experience over a window. In default, it uses a prediction from the stable learner. If a performance of the reactive learner exceeds the one of the stable learner at certain level, the paired learner swaps to use a prediction of the reactive learner. Similarly, Nishida and Yamauchi [16] use paired learners, the long window size and the short window size. They swap the usage of a learner based on statistical difference tests. Kuncheva and Žliobaitė [11] dynamically changes the window size according to the statistical test as well. While an advantage of the statistical test is a proposal of a theoretically accurate window size, the number of observations is required to be accurate.

Our proposing method contains paired evaluators similarly to paired learners. A key difference is that we do not adjust the window size of forecast. In stead of setting appropriate window size of forecast, evaluators in our method select an appropriate base forecast having a certain window size by evaluators. Similarly to FLORA and paired learners, our method corresponds to recurring contexts by learning from the past evaluation errors.

3 Model

In this section, we detail our proposing forecast method, paired evaluators method (PEM). In our method, there is a set of base forecasts that generate forecast values independently. We used paired evaluators to evaluate forecast accuracy in different ways. We evaluate the performance of evaluators themselves according to the historical data. The selection of an evaluator is based on learning-algorithm. Our method is one of heuristics that does not require many observations.

3.1 Preliminaries

Let $\mathcal{T} = \{-T, \ldots, -1, 0, 1, \ldots, T\}$ be the set of discrete time and $t \in \mathcal{T}$ be a certain time. We call $t = 0$ as the current time. We denote $\mathcal{H} = \{t \in \mathcal{T} : t < 0\}$ as the set of historical periods. Let $X(t) \in \Re^z$ be a vector in z-dimensional feature space observed at time $t \in \mathcal{T}$ and $y(t) \in \Re$ be its corresponding label to be predicted. We refer to $X(t)$ as an instance, a pair $(X(t), y(t))$ as a labeled instance, instances $(X(-T), \ldots, X(-1))$ as historical data X^H, and an instance $X(0)$ as a target instance. As time is incremented, the number of historical data is increased and the current time is shifted. Notice that a target instance $X(0)$ is not observed until time is incremented.

3.2 An Overview of Our Proposing Forecasting Method

In this section, we present an overview of our forecasting method. We use several different base forecasts and select a base forecast which is expected to be the best forecast according to the past experiences. We refer a way to select a base forecast as an evaluator. A key characteristic of our method is an evaluation of performances of evaluators.

Let f be a base forecast. Let $\mathcal{F}$ be a set of base forecast. Let us denote f^i as the i-th forecast among $\mathcal{F}$ and we also denote $\mathcal{I}$ as the set of forecasts. We denote $\delta^i(t) = y(t) - f^i(X^H(t))$ as a forecast error at time t which is obtained at time $t+1$.

Our proposing forecast method proposes the optimal base forecast $f^{\hat{i}}$ among a set of base forecasts $\mathcal{F}$ without severe parameter tunings. As depicted in Figure 1, this method consists of four parts: (1) pre-processing, (2) base forecast proposal according to evaluator, (3) evaluator selection, and (4) forecasting. In part (1), past forecast errors δ for all base forecast are calculated, respectively.

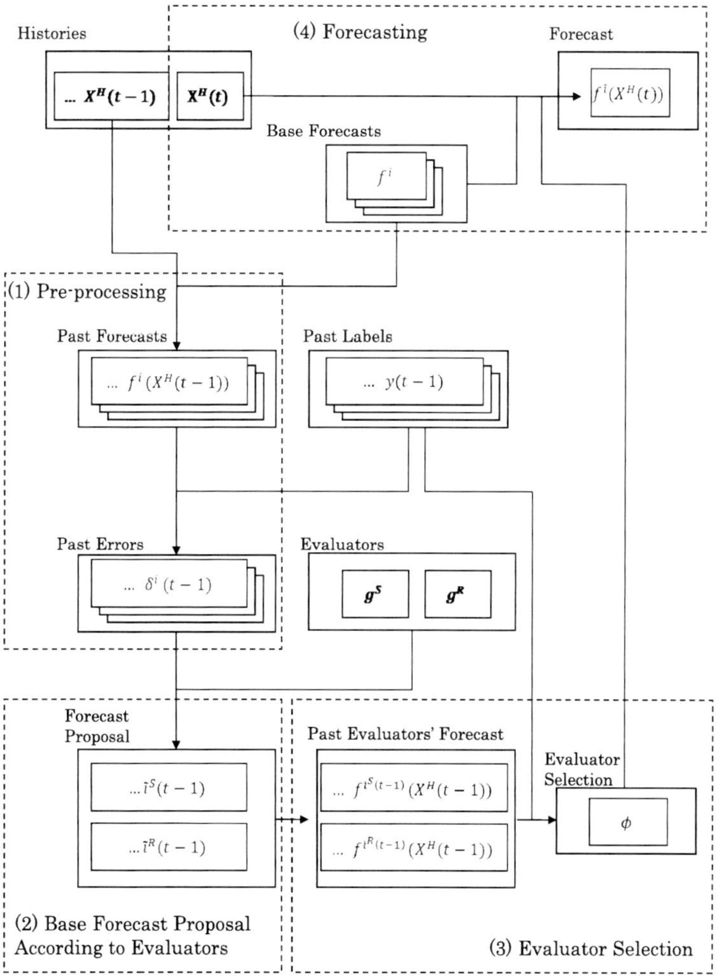

Fig. 1. Structure of Paired Evaluator Method

In part (2), we use paired evaluators that are a stable evaluator and a reactive evaluator. Each evaluator proposes the expected best base forecast based on different weights for evaluations. Let $\mathbf{w}$ be a weight vector of forecast errors. Let $\tilde{i}$ be the $\tilde{i}$-th base forecast which is expected to be the best base forecast. Multiplying a weight vector to forecast errors, we are able to compare performances of base forecasts and we are also able to obtain the estimated best base forecast as follows:

$$\tilde{i} = \arg\min_{i \in \mathcal{I}} \sum_{t \in \mathcal{H}} w(t)\delta^i(t) \tag{1}$$

where $w(t)$ is an element of the weight vector. According to Equation (1), a base forecast that minimizes the weighted errors is estimated as the best base forecast. Here, let us denote g as an evaluator that selects the optimal base forecast based on Equation (1). We present an example of paired evaluators in Figure 2. If a weighting vector has heavier weights for more recent errors, it prefers short-term forecast accuracy which is a reactive evaluator g^R. Contrary, the flat weight vector prefers long-term forecast accuracy which is a stable evaluator g^S. Long-term accuracy is preferred in general. However, right after a concept drift, the long history of forecast errors may not tell a proper forecast. According to these errors and evaluators, a base forecast is proposed as shown in Equation (1). Let $\tilde{i}^S(t)$ be the expected best base forecast according to the stable evaluator which is $\tilde{i}^S(t) = g^S(\boldsymbol{\delta}^H(t))$, where $\boldsymbol{\delta} = \{\delta^i\}_{i \in \mathcal{I}}$. Similarly, the expected best base forecast according to the reactive evaluator is $\tilde{i}^R(t) = g^R(\boldsymbol{\delta}^H(t))$.

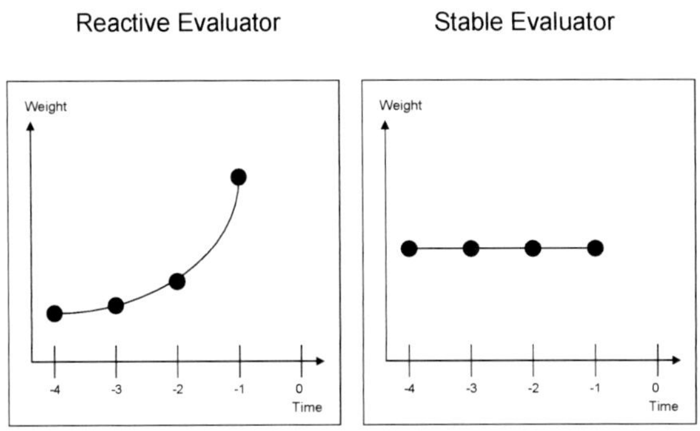

Fig. 2. Examples of Evaluators

Since the proposed base forecast depends on the setting of the evaluator, a central issue is how to select evaluators which is part (3) of our method. With respect to the selection, there are some key ideas behind our method. First, the stable evaluator works well under versatile situations. It is used as a default evaluator. Second, the stable evaluator may not work well right after a poor performance. If the reactive evaluator has performed better than the stable evaluator at the similar cases in the past, our method switches to use the reactive evaluator. Finally, if past experiences are inconsistent, recent experiences have a greater importance for decision makings. Considering these aspect, we select an evaluator based on a learning-algorithm. Let Φ be an evaluation function of evaluators that assigns a degree of superior evaluator on a performance of the stable evaluator. Based on this evaluation function, we obtain the expected best base forecast $\hat{i}$ such that:

$$\hat{i}(t) = \begin{cases} \tilde{i}^S(t), & \text{if } \Phi(t; \delta^{\tilde{i}^S(t-1)}(t-1)) \geq \theta; \\ \tilde{i}^R(t), & \text{otherwise.} \end{cases} \tag{2}$$

where θ is a threshold parameter. We detail how the evaluation function is updated according to the past experiences in the following section.

Once an evaluator is selected in part (3), part (4) is directly induced and we obtain the best performing base forecast $f^{\hat{i}}$ and its forecast value $f^{\hat{i}}(X^H(t))$. In the following section, we detail part (3) of our proposing forecasting method.

3.3 Update Rules of the Evaluation Function

The evaluation function of evaluators is updated according to a learning-based approach that consists of three types of update rules: (i) initialization, (ii) a performance of an evaluator exceeded a performance of another evaluator, and (iii) no differences on performances between two evaluators. Regarding the first update rule, we set $\Phi(\delta) = 0$ for all δ. The second rule is for cases where one evaluator performs better than another. Once actual value $y(t-1)$ is realized at time t, we obtain forecast errors of the stable evaluator and the reactive evaluator, $\delta^{\tilde{i}^S(t-1)}(t-1)$ and $\delta^{\tilde{i}^R(t-1)}(t-1)$, respectively. If the stable evaluator performs better than the reactive evaluator, i.e., $\delta^{\tilde{i}^S(t-1)}(t-1) < \delta^{\tilde{i}^R(t-1)}(t-1)$, $\Phi(t;\delta) := \Phi(t;\delta) + \lambda$ for all $\delta \leq \delta^{\tilde{i}^S(t-1)}(t-1)$, where $\lambda > 0$ is an update coefficient. Contrary, if the stable evaluator performs worse than the reactive evaluator, i.e., $\delta^{\tilde{i}^S(t-1)}(t-1) > \delta^{\tilde{i}^R(t-1)}(t-1)$, $\Phi(t;\delta) := \Phi(t;\delta) - \lambda$ for all $\delta \geq \delta^{\tilde{i}^S(t-1)}(t-1)$. The third rule is for cases where both evaluators perform equal. In such cases, the effects of past experiences are reduced by a reducing coefficient $0 \leq \alpha \leq 1$ as follows, $\Phi(t;\delta) := \alpha\Phi(t;\delta)$ for all δ.

4 Paired Evaluators Method and Drift Types

In this section, we show some simple examples in order to show how our proposing forecasting method, paired evaluators method (PEM), deals with typical drift types, sudden drift, incremental drift, gradual drift, and recurring contexts. Notice that a single window size approach has a problem to deal with these drift types. While smaller window sizes tend to fit for sudden drift and incremental drift, they are too sensitive for gradual drift and recurring contexts. It is significant to switch to an appropriate window size that corresponds to an observed drift type. In order to switch the window size, PEM uses two types of evaluators, the stable evaluator and the reactive evaluator, as we have shown in the previous section. This approach, particularly, works for gradual drift and recurring contexts as we show in some examples.

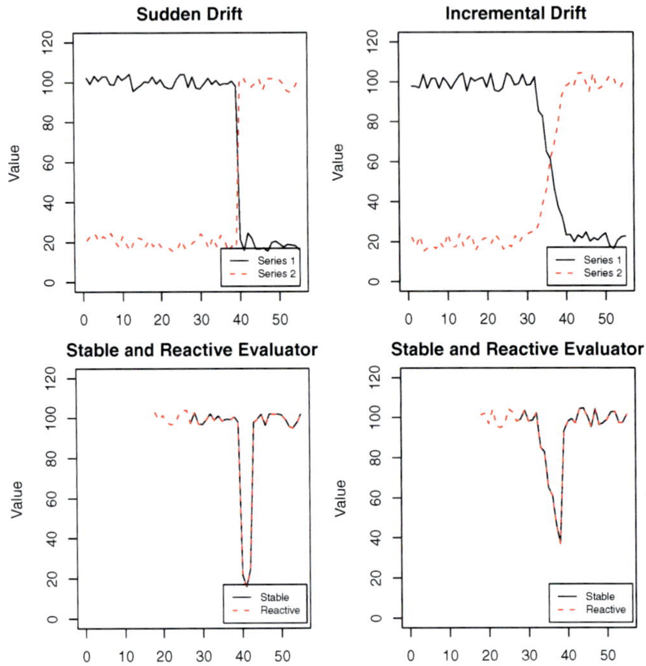

Fig. 3. Sudden Drift and Incremental Drift

We prepare the set of artificial data that characterizes typical drift types. There are two time series that are either $100 \pm \epsilon$ or $20 \pm \epsilon$ where ϵ is randomly drawn from uniform distribution ranging $[-5, 5]$. A key task is to predict a series that will be the greater value in the following time. Regarding PEM, we use three base forecasts, 3-month average, 6-month average and 12-month average; we set an update coefficient as 1, a decreasing coefficient as 0, a threshold as 0. As for parameter of evaluators, we use 12 months equitable weights $w = \{1, 1, 1, 1, 1, 1, 1, 1, 1, 1, 1, 1\}$ for a stable evaluator and 3 months equitable weights $w = \{1, 1, 1\}$ for a reactive evaluator.

At first, we show examples of sudden drift and incremental drift in Figure 3. The top graphs show series 1 and 2 of the respective drift patterns. For both cases, series 1 is to be predicted at the beginning, series 2 swaps at certain time, and series 2 is to be predicted after the swap. While sudden drift occurs at time 39, incremental drift starts changing at time 33, takes over observed at time 37, and ends changing at time 39. The bottom graphs show performances of paired evaluators. For both drift patterns, both evaluators have exactly the same performances. They require the minimum time, three time periods, to correspond to the drifts. Differences between the stable evaluator and the reactive evaluator are observed if drifts occur more frequently.

Next, we show examples of gradual drift and recurring contexts in Figure 4. The top graphs show series 1 and 2. For both cases, series 1 is to be predicted at the beginning and sudden changes occur frequently. While occurrence of swaps becomes more

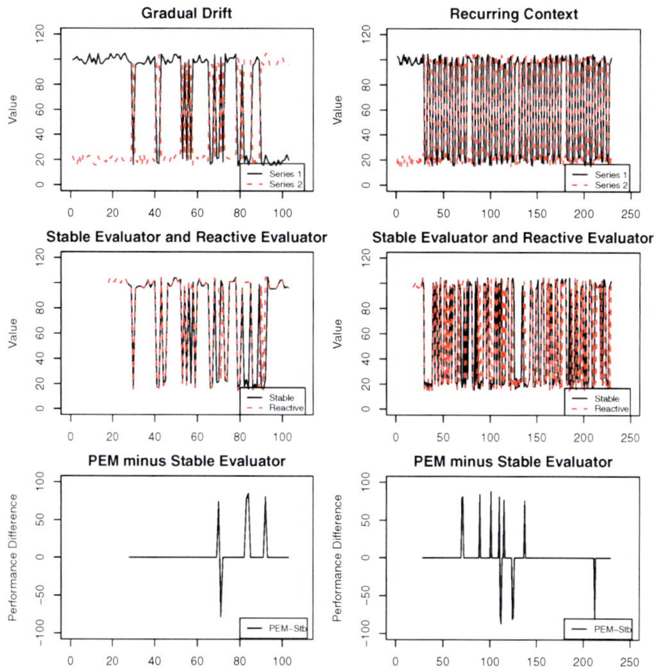

Fig. 4. Gradual Drift and Recurring Contexts

frequent over time in gradual drift, swaps occur cyclically and randomly in recurring contexts. In such cases, some inconsistencies of past experiences occur between the stable evaluator and the reactive evaluator as we show in the middle of the graphs. In some cases, the past experience works for a good forecast. If the past experiences work good, PEM tends to follow such experiences. In the bottom graphs, we show difference of performances of PEM and the stable evaluator. PEM tends to exceed the performance of the stable evaluator if one evaluator works better than another in consecutive times.

In reality, it is difficult to know or identify drift patterns before occurrences. In the following section, we use actual financial data to show performances of PEM considering such difficulties.

5 An Application Example - Financial Data

In this section, we show an application example using financial data. One of the main interests in finance is dealing with time, price and risk due to difficulties of precise forecasts. We view sudden and sharp market changes as concept drift problems. Sharpe [18], Lintner [13] and Mossin [14] are main contributors of an establishment of capital asset pricing model (CAPM) that has played a key role in modern portfolio theory. A main characteristic of the CAPM is an emphasis on a price discovery process rather than pricing itself. In the CAPM, a theoretically appropriate required rate of return of an asset is obtained according to a consideration of the expected return of the market, the expected return of a theoretical risk-free asset and non-diversifiable risk. Hence, the non-diversifiable risk is used as a single *factor* to compare the excess returns of a portfolio with the excess returns of the entire market that entails the set of optimal equities for a portfolio. In order to reduce some exceptional cases of the CAPM, Fama and French [6] add two risk factors, *value* and *size*, and Carhart [4] adds a factor, *momentum*. Nowadays, these models are widely used as fundamental models to explain stock price movements.

A key issue with respect to multiple factors model is a selection of the most effective factor. Difficulties to predict the most effective factors are two-folds. First, the price discovery process is not consistent over time for each factor. Some factors may be rapidly effective, but others may take time. It is important to find the factor that is effective immediately. Second, an effective factor can be either forward or contrary. While a large absolute value can be an effective factor, it can be a risky factor if it changes its sign frequently. Hence, a main objective of the forecast is to predict both the factor that has the largest absolute value and its sign. For this objective, we apply our proposing forecasting method under dynamically changing market environments. We remark that these factors are explained to work for the price discovery process in Finance. Hence, our experiment is not just find opportunities of investments from by chance. Instead, we investigate how a key factor of the price discovery process changes over time. In the following section, we describe the experimental setups.

5.1 Experimental Setups

We conduct an experiment based on a back-testing using historical monthly factor data of US equities which can be obtained from [7]. In the back-testing, we apply our proposing forecasting method to the historical data and compare performances with other forecasting methods. We use the following three factors classified by 10 percentile from 1964 to 2009 that have already mentioned above:

- momentum: historical price increase for 12 months,
- value: book-to-market ratio,
- size: size of a firm regarding market capitalization.

It is common to investment on spreads between equities in the top group and the bottom group of the effective factors. The respective spreads are known as *winner-minus-loser* for momentum, *high-minus-low* for value and *big-minus-small* for size. Among the 10 percentile data, we focus on factor spread between the top 10 % and the bottom 10 %. If one factor is positively effective, it is common for hedge fund operations to take a combination of purchase of equities in the top 10 % and short-sell of equities in the bottom 10 %. Behind these operations, there are two practical reasons: (1) Equities in the two extremes should represent behaviors of factors most clearly, and (2) Investments of the both sides should reduce risks.

In the experiment, we allow our proposing forecasting method to make an investment for 1 unit for a single factor in each month including contrarian trading. Therefore, a key task is to choose one action from 6 candidates (3 factors $\times$ $\{$follow or contrary$\}$) that gains the maximum profit in the following month constantly.

5.2 Basic Statistics of Factor Data

Prior to show the experimental result, we briefly show some characteristics of raw data used for the back-testing. In Table 1, we present basic statistics of three types of factor data classified by decades. Average stands for actual performances of each factor. Notice that a negative sign is desired for the size factor, since smaller size companies are expected to grow faster as proposed in Fama and French [5]. According to the basic statistics, momentum has gained great performances until 1990s; however it drops sharply in 2000s. A standard deviation of momentum in 2000s is extremely high compared to the ones in different decades and the ones of the different factors. As a standard deviation per average indicates, investments on momentum in 2000s are risky. A significant difference in 2000s is due to occurrences of some liquidity shocks caused by the September 11 attacks in 2001, speculation on natural resources relevant to the 2003 invasion of Iraq, the end of the monetary easing policy in 2004, and the late-2000s recession.

Notice that key points of investments are not only limited to detect factors that are constantly increasing or decreasing. Moreover, if a forecasting method is able to predict efficient factors in changing environments, it is considered as a significant method. In order to clarify target durations, we present a benchmark that is average of the three factors, $(momentum + value - size)/3$. In Figure 5, we present cumulative factor spread returns of three factors classified by decades and the average of these three factors. We remark that we consider a negative for the size factor. The average of the three factors indicates that the early 1970s, the late 1990s and the late 2000s are difficult durations to gain high-performance. We evaluate the performance of our proposing forecast model focusing on these periods.

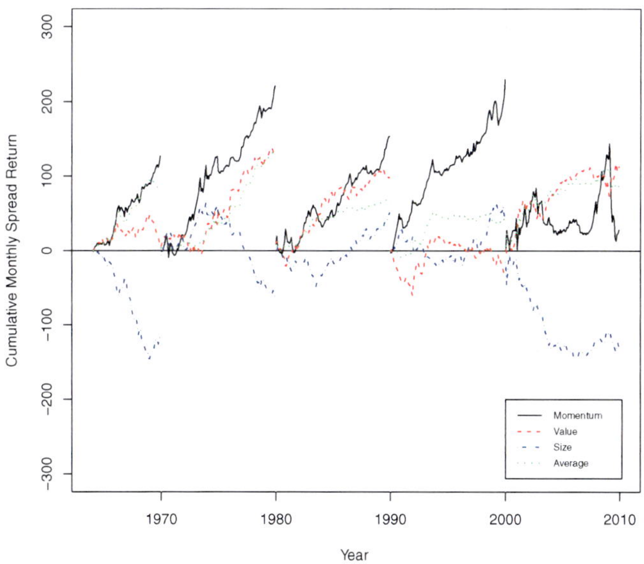

Fig. 5. Cumulative Monthly 10 Percentile Spread Return of Factor Investments Classified by Decades

Table 1. Basic Statistics of Factors

	Momentum			Value			Size		
	Ave	SD	SD/Ave	Ave	SD	SD/Ave	Ave	SD	SD/Ave
1960s	1.76	4.40	2.49	0.09	3.77	38.08	-1.57	5.11	-3.24
1970s	1.84	6.22	3.37	1.09	5.15	4.70	-0.52	5.48	-10.42
1980s	1.27	5.18	4.06	0.82	4.04	4.90	0.42	3.71	8.68
1990s	1.91	5.75	3.00	-0.30	4.46	-14.79	0.36	4.49	12.43
2000s	0.23	11.23	47.30	0.94	5.34	5.64	-1.07	5.82	-5.40
Overall	1.37	7.17	5.21	0.57	4.67	8.20	-0.38	5.00	-13.09

5.3 Experimental Results

In this section, we compare the three forecasting models based on a back-testing using Fama-French financial data. First, we describe some configurations of parameters used for PEM and two well-known dynamic window size methods, CWA and FLORA. Then, we show experimental results.

With respect to PEM, we use 6 basic forecasts, 3-month top mode, 6-month, 9-month, 12-month, 18-month, 24-month average. 3-month top mode selects the most frequent factor that performs the best among three months. We use smoothed data for averages in which highly effective points exceeding 1.5 standard deviations are reduced. As for parameter of evaluators, we use 12-month equitable weights $w = \{1,1,1,1,1,1,1,1,1,1,1,1\}$ for a stable evaluator and 5-month decreasing weights $w = \{0.03125, 0.0625, 0.125, 0.25, 0.5\}$. With respect to coefficients, we set an update coefficient as 1, a decreasing coefficient as 0.005, a threshold as 0.

Regarding CWA, there are three types of windows size, small-medium-large, with default window sizes, 3, 6 and 12-month, respectively. If a distance between the latest instance and the historical instances is within a consistent coefficient 3.88, which is 2 standard deviation of the sum of the absolute instances within the first 12 months, the medium window size is enlarged up to 11-month and the large window size is the double size of the medium window size. If consistencies are persistent more than 12 months, the large window corresponds to the size of persistence. According to averages of three window sizes, the best factor is calculated. The type of the window size is selected according to the performance of the previous month.

With respect to FLORA, we tuned the algorithm in order to deal with our test data as follows. We classify instances into 24 states that consist of the best factor and the second best factor with signs. The state patterns are to allocate the following two { the best factor with a positive sign, the best factor with a negative sign } and { the second best factor with a positive sign, the second best factor with a negative sign } in the three positions that are $2 \times 2 \times 2 \times 3$. Once FLORA observes a new instance, it looks up the same state in the past. Among the matched state, it calculates the most frequent top factor in the following month which is used for a prediction in this month. We set default search periods as 36 months. If the matched state is less than a minimum number of match 5, it grows the number of windows up to 48 months. If the accuracy is greater than $50\ \%$ and the recent result is inaccurate, reduces window size $20\ \%$ where the minimum window size is 24 months. Otherwise, keep the same window size as the previous month.

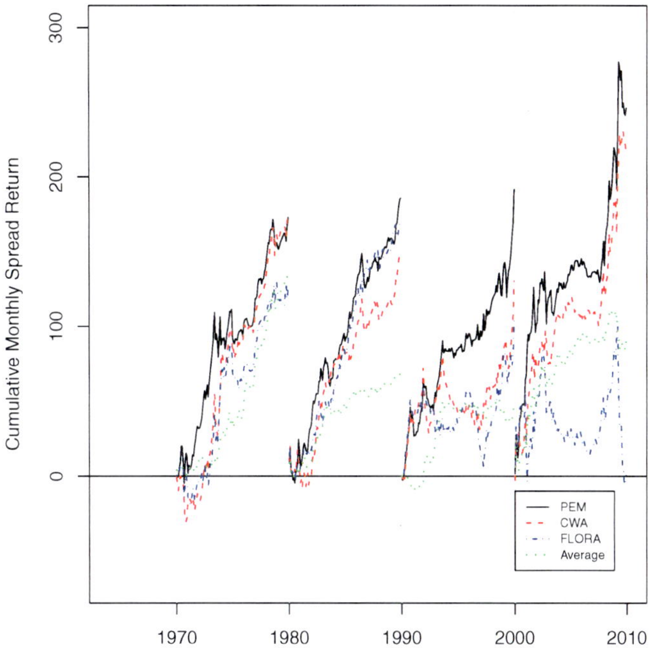

Fig. 6. Forecast Performances on Cumulative Monthly 10 Percentile Spread Return of Portfolios Classified by Decades

Now, we show the experimental results. We depict the cumulative performances of the three models classified by decades in Figure 6. According to the experiment, PEM performs better than the other two models in most of time. An important aspect to evaluate performances is persistency of growth. In most time, PEM continuously performs well. While 1970s, the early 1980s, the early 1990s, and the early 2000s are relatively easy periods according to the average performance, PEM is quite stable. Even though the rest of periods are not easy, it performs good due switches of evaluators effectively during this periods. Surprisingly, growth during this period is quite remarkable that includes the financial crisis of 2008-2009. Key reasons of this are (i) PEM quickly adapt to the drift which is the reverse of momentum, and (ii) This works consecutively.

CWA also performs well similarly to PEM. However, in the middle of 1990s, it looses its control for a while. A disadvantage of CWA is a change of window size is one even though they have three sizes of windows. Hence, it may take time to search an appropriate window size. With respect to FLORA, performances are not good dynamically changing environments such as the middle of 1990s and 2000s. The performance of FLORA is good if there are many sample data in the past. However, in our data set, this is not always true. Similarly, paired learner in [1,16] is not implemented in this experiment, since a reactive learner is not efficient with a small number of reference periods with our data.

6 Conclusions

In this paper, we have introduced our forecasting method, paired evaluators method, which tends to improve forecast performances for consecutive concept drift patterns, such as gradual drift and recurring contexts. In our method, a set of base forecasts is used for a prediction which is selected by evaluators. We use paired evaluators, a stable evaluator and a reactive evaluator. A selection of evaluator is based on learning algorithm which learns the past performances of evaluators. By learning, paired evaluators method continuously attempts to detect an alternative evaluator to improve forecast accuracy. This approach suits for consecutive concept drift patterns.

We have introduced a methodology to deal with a financial investment problem, a factor selection problem, using concept drift solutions. Experimental results show that our proposing method has discovered effective factors more efficiently than the other two representative methods, CWA and FLORA, which change the forecast window size dynamically. Our method is robust against many difficult circumstances including the late-2000s recession. Moreover, a similar performance is confirmed in Japanese Financial data [19].

In a broader sense, paired learners for online classifications, such as paired learners having two different window sizes for classifications based on naive Bayes approach [1] and Todi (two online classifiers system for learning and detecting concept drift) based on a statistic test [16], are similar to our approach. They use paired classifiers to control stability and reactivity for changes over time. While they directly set the window size on classifiers, which is fixed, in our approach, window sizes are set by respective base forecasts instead of evaluators. Hence, our approach uses multiple window sizes for a set of forecasts and a forecast having the most appropriate window size tends to be selected by evaluators. In our experiments, these paired learners for classifications are not used, since our experimental data does not have similar features in short periods.

In the financial investment problem, we have focused on a selection of the best factor. This is not a restriction in practice. In the future, we consider the method to set the most appropriate weights for the three factors.

References

1. Bach, S., Maloof, M.: Paired learners for concept drift. In: Proceedings of the 8th IEEE International Conference on Data Mining, pp. 23–32 (2008)
2. Bifet, A., Gavalda, R.: Learning from time-changing data with adaptive windowing. In: Proceedings of SIAM International Conference on Data Mining (SDM 20007), pp. 443–448 (2007)
3. Bifet, A., Holmes, G., Pfahringer, B., Kirkby, R., Gavalda, R.: New ensemble methods for evolving data streams. In: Proceedings of the 15th ACM SIGKDD International Conference on Knowledge Discovery and Data Mining, pp. 139–148 (2009)
4. Carhart, M.M.: On persistence in mutual fund performance. Journal of Finance 52(1), 57–82 (1997)
5. Fama, E.F., French, K.R.: The cross-section of expected stock returns. Journal of Finance 47(2), 427–465 (1992)
6. Fama, E.F., French, K.R.: Common risk factors in the returns on stocks and bonds. Journal of Financial Economics 33(1), 3–56 (1993)

7. French, K.R.: Fama/french factors in u.s. research returns data, http://mba.tuck.dartmouth.edu/pages/faculty/ken.french/data_library.html (accessed July 23, 2010)
8. Gama, J., Medas, P., Castillo, G., Rodrigues, P.: Learning with Drift Detection. In: Bazzan, A.L.C., Labidi, S. (eds.) SBIA 2004. LNCS (LNAI), vol. 3171, pp. 286–295. Springer, Heidelberg (2004)
9. Giacomini, R., Rossi, B.: Detecting and predicting forecast breakdowns. Review of Economic Studies 76(2), 669–705 (2009)
10. Kolter, J., Maloof, M.: Dynamic weighted majority: An ensemble method for drifting concepts. Journal of Machine Learning Research 8, 2755–2790 (2007)
11. Kuncheva, L.I., Žliobaitė, I.: On the window size for classification in changing environments. Intelligent Data Analysis 13(6), 861–872 (2009)
12. Lazarescu, M.M., Venkatesh, S., Bui, H.H.: Using multiple windows to track concept drift. Intelligent Data Analysis 8(1), 29–59 (2004)
13. Lintner, J.: The valuation of risk assets and the selection of risky investments in stock portfolios and capital budgets. Review of Economics and Statistics 47(1), 13–37 (1965)
14. Mossin, J.: Equilibrium in a capital asset market. Econometrica 34(4), 768–783 (1966)
15. Nara, Y., Ohsawa, Y.: Tools for shifting human context into disasters: a case-based guideline for computer-aided earthquake proofs. In: Proceedings of the 4th International Conference on Knowledge-Based Intelligent Engineering Systems and Allied Technologies, pp. 655–658 (2000)
16. Nisida, K., Yamauchi, K.: Learning and detecting concept drift with two online classifiers. In: Proceedings of the 22nd Annual Conference of the Japanese Society for Artificial Intelligence, pp. 3C2–1 (2008)
17. Ohsawa, Y., Nara, Y.: Decision process modeling across internet and real world by double helical model of chance discovery. New Generation Computing 21, 109–121 (2003)
18. Sharpe, W.F.: Capital asset prices: A theory of market equilibrium under conditions of risk. Journal of Finance 19(3), 425–442 (1964)
19. So, J., Furuhata, M., Mizuta, T.: Operational model considering transaction costs to correspond to sudden changes in japanese stock markets. In: Proceedings of the 6th workshop SIG-FIN in the Japanese Society for Artificial Intelligence, pp. 23–29 (2010)
20. Žliobaitė, I.: Learning under concept drift: an overview. Technical report, Vilnius University, Faculty of Mathematics and Informatics (2009)
21. Widmer, G., Kubat, M.: Learning in the presence of concept drift and hidden contexts. Machine Learning 23(1), 69–101 (1996)

Efficient Service Discovery among Heterogeneous Agents Using a Novel Agent Ranking Algorithm

Hao Lan Zhang[1], Clement H.C. Leung[2], Chaoyi Pang[3], and Xijin Tang[4]

[1] IS Discipline, NIT, Zhejiang University, Ningbo, China
haolan.zhang@nit.zju.edu.cn
[2] Hong Kong Baptist University, Hong Kong
clement@comp.hkbu.edu.hk
[3] The Australian E-health Research Centre, CSIRO, Australia
Chaoyi.Pang@csiro.au
[4] Chinese Academy of Science, Beijing, China
xjtang@amss.ac.cn

Abstract. Discovering and matching services among heterogeneous agents is the fundamental process for agent cooperation and coordination in multi-agent systems. In such a process, numerous service-providing agents are matched against requested criteria. During the matching procedure, an agent is given the chance to discover its new corresponding agents that can provide desirable services. However, in order to solve more complex problems, the number of agents in a multi-agent system is increasing rapidly; therefore making the chance for an efficient and desirable match is decreased. Thus, improving the efficiency of the agent matching process has become an important issue in the multi-agent filed. Utilising an appropriate agent-matching mechanism will enhance agent cooperation and communication efficiency in dealing with complex problems. In this chapter, we develop a new agent-matching algorithm, the Agent-Rank (AR hereafter) algorithm, which ranks service-providing agents according to their contributions to a nominated requesting agent. The AR algorithm improves agent matching process through combining the general ranking scores with the request-based ranking scores. The AR algorithm narrows down the searching targets (services), which is avail to detect some random targets in a chance discovery process.

Keywords: Service discovery, agent matching, ranking algorithm, Multi-agents.

1 Introduction

Matching an agent's request against other agents' services to solve a complex problem cooperatively is a common method used in multi-agent systems. In such matching processes, the agent sends requests for matching is called Requesting Agent (R-agent hereafter) and the agent provides services to the requesting

Y. Ohsawa & A. Abe (Eds.): Advances in Chance Discovery, SCI 423, pp. 143–161.
springerlink.com

agent is called Service-providing Agent(P-agent hereafter). Multi-agent systems can solve many complex problems by incorporating various task-oriented agents. For instance, a complex financial investment plan can be decomposed into numerous subtasks or procedures such as an investment policy, financial forecasts, organisational resources, credit evaluation and so on. A task-oriented agent can represent as a problem-solver in each subtask domain. Thus, a complex problem can be tackled with high degree of efficiency and flexibility. Applying an effective agent cooperation and communication process is the key to a multi-agent system since no agent possesses a global view of the entire system [1]. However, the chance for a successful match between an R-agent and a P-agent has been affected by the increasing number of agents. A multi-agent system must provide some competent agent matching and searching mechanisms in order to improve the cooperation and communication efficiency among agents; and it also should enable agents with a local view of their environment to take action that advance global system objectives [2].

Existing agent-matching mechanisms are helpful for improving the efficiency of interaction and cooperation between agents and web services [3,4,5]. Nevertheless, current knowledge-based reasoning mechanisms are facing challenges in providing an efficient semantic match based on different information resources [6,7]. This affects the quality and accuracy of an agent matching process. We often experience the difficulties in obtaining accurate information in our day-to-day life. For instance, Web search engines give us some irrelevant information; language translation software misunderstands our words, etc. Therefore, the agent-matching process has to be highly tolerant and can specifically measure services against requests. The expectation of exact matches in every matching process is not realistic. However, narrowing down the matching objects in a complex matching process can significantly improve matching efficiency.

A new agent-matching algorithm is introduced in this paper, namely the AR algorithm. The use of the AR algorithm can enhance the efficiency of the existing agent-matching mechanisms. The AR algorithm is inspired by the *PageRank* algorithm [8,9], which has been successfully applied in Web search applications.

The AR algorithm is based on a multi-agent framework [10,11]. The algorithm can be further used in other agent-based systems through adopting a complex graph-based agent association analysis (see Section 3). The primary concept of the AR algorithm is to rank all the P-agents in a multi-agent network. According to the ranking information, an agent-matching process is able to select appropriate P-agent(s) more efficiently and accurately. Normally, an agent-matching process selects the highest rank P-agent.

The AR algorithm supports both human-based and agent-based decision-making processes. The AR algorithm allows a decision- maker (human or agent) to choose some of the most relevant P-agents according to their ranks. Existing agent-matching mechanisms can improve their matching efficiency and quality through adopting the AR algorithm in their matching processes. Thus, the current matching processes will search for P-agents according to the provided ranks instead of massive searching.

This chapter is organised as follows. The next section reviews several major agent matching methods and their related work. Section 3 and 4 introduces the general-ranking algorithm and the request-based ranking algorithm, respectively. The last section concludes the research findings.

2 Related Work

Studies on agent matching mechanisms have drawn much attention of researchers from the multi-agent field. Several matching models have been built to improve the matching efficiency, such as the two-sided search model [12], similarity measure model [13] based on Stackelberg equilibrium theory, etc. Most of existing agent-matching methods deploy description language-based matching, such as LARKS (Language for Advertisement and Request for Knowledge Sharing) [4], SDL (Service Description Language) [14], etc. These methods mainly rely on matching agent-service-descriptions, also called service advertisements. The descriptions indicate an agent's capabilities and services. An R-agent can search for a P-agent through matching the descriptions provided by the other agents. In LARKS, there are three matching mechanisms [4]:

Exact Match: This is the most accurate matching, which requires that both service descriptions being literally equal.

Plug-In Match: A less accurate matching method in LARKS, in which two matching descriptions are generally similar. Exact match is a special case of plug-in match.

Relax Match: The least accurate matching method. It determines the similarity of two descriptions by returning a numerical distance value instead of semantically matching the two descriptions.

All of these matching mechanisms are essentially based on semantic matching, which involves searching an appropriate agent from their information providers. These mechanisms often neglect the probabilistic factors in the matching process. For instance, a plug-in match process might match an appropriate agent from a large multi-agent system. In fact, thousands of agents can fulfil the search requirement and the first matched agent might not be the most appropriate P-agent. Unfortunately, current agent- matching mechanisms do not consider this circumstance and this increases the failures in current agent matching processes. Moreover, current agent-matching mechanisms often neglect some important factors, such as response quality, visiting history, etc. This flaw affects the effectiveness of an agent-matching process. For instance, an unpopular agent might be constantly selected as a P-agent just because it meets the semantic description but actually it is not the appropriate agent or even the knowledge of the selected agent is out of date. The use of the AR algorithm could eliminate these problems by incorporating various factors that affect agent matching efficiency in the ranking process. The AR algorithm is a method to improve the efficiency of existing agent-matching mechanisms. In particular, the re-matching process is highly efficient in AR-based systems. For instance, when the selected P-agent(s) is (are) inappropriate, the re-match process is much easier than in current mechanisms. The re-match process in the AR algorithm does not require searching the agent information server again. It simply continues its operation based on the ranking list generated previously.

The AR algorithm adopts the underlying concept of the PageRank algorithm [8,9] but it has its own ranking mechanisms. First, PageRank is a Web page-based mechanism, whereas the AR algorithm deals with the agent-matching process within multi-agent-based systems. Second, PageRank does not consider the response quality issue in its matching process. On the contrary, the response quality factor is one of the major factors that affect an agent's ranking score in the AR algorithm. Third, PageRank forms a probability distribution over Web pages. Unlike PageRank, the AR algorithm indicates the ranks of the P-agent candidates associated with an R-agent. In other words, the frame of reference in PageRank is an overall network, whereas it is a specific R-agent in the AR algorithm.

A comparative ranking function has been suggested in [15]. However, the comparative ranking function mainly deals with goal-oriented tasks and neglects some important factors in agent cooperation such as response quality, visit frequency, agent importance, etc. Moreover, this early study [15] does not distinguish between the P-agent and R-agent, which complicates the ranking process because the ranking objects involve all agents in a system instead of a group of related P-agent candidates. This results in the comparative ranking process becoming extremely complicated when a large number of agents are involved.

3 The AR Algorithm-Part I: General Ranking

There are three major stages in an agent-ranking process. The first stage is the General ranking process, which ranks the importance of all the agents that have relationships with the R-agent. The second stage is the Request-based ranking process, which specifically focuses on specific searching and matching based on a request and incorporates other major factors such as visiting history, quality of service and the results from the general ranking stage.

The general ranking process ranks the importance of all the agents related to the R-agents. The related agents refer to those agents that send or receive requests from or to the request agent. The general ranking process adopts the basic idea of PageRank that calculates agents' incoming and outgoing links to form an importance distribution over an agent network. Unlike PageRank, the visit frequency of an R-agent to a P-agent is calculated in the general ranking process because the frequency of usage of a P-agent is an important criterion for evaluating an agent's importance. We use an Agent Association Graph (AAG), shown in Fig.1, to help analyse the ranking process. The AAG represents the cooperation history of an R-agent with its associated agents. Drawing an AAG is always the first step in analysing the general ranking process. Fig.1 illustrates an example to explain the general ranking process. Fig.1 is a Small world topology graph [16] formed by an R-agent and its associated agents (P-agent candidates). Here, R-A denotes the R-agent.

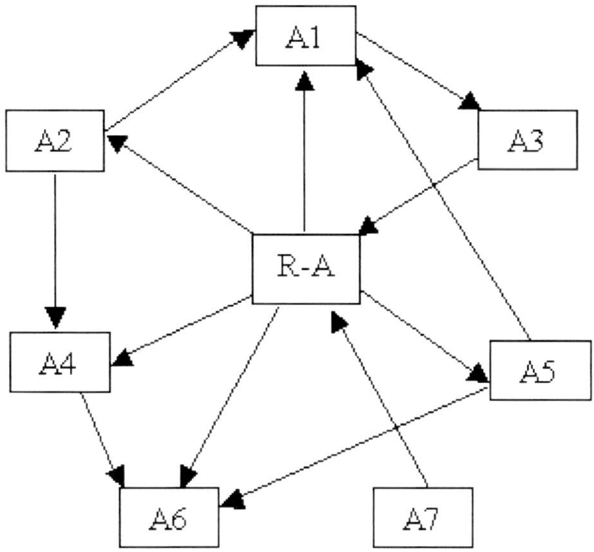

Fig. 1. An Agent Association Graph

A large scale free agent system (as shown in Fig.2) can be viewed as the form of numerous AAGs. In each AAG, the R-agent (shown as "R-A" in Fig.1) is the centre of the graph. The links in these graphs represent previous cooperation between the R-agent and the other agents. The general ranking process ranks all the P-agent candidates in the R-agent-based graph. The ranking process is based on the calculation of the total inbound links of the P-agents. The inbound links of an agent in an AAG are those arrows directing at this agent; and the outbound links of an agent are those arrows originating from this agent.

3.1 Factors in the General Ranking Process

There are several crucial details that need to be considered including the total number of links in a graph, the damping factor, the visit frequency by the R-agent, the total visit number in a period, and the response quality. The response quality factor concerns the response speed (time) of a P-agent to an R-agent. The response time is a key factor affects the response quality. The total number of links, which can be systematically evaluated node by node:

$$E = \frac{\sum_{i=1}^{m} G_i + C_i}{2} \quad (1)$$

where E denotes the total number of links in an AAG; G_i denotes the outbound links of an agent; C_i denotes the inbound links of an agent in an AAG; and m denotes the total agent number, including the R-agent, in an AAG. The reason for Eq. 1 is that: the inbound links of each node (agent) are added to

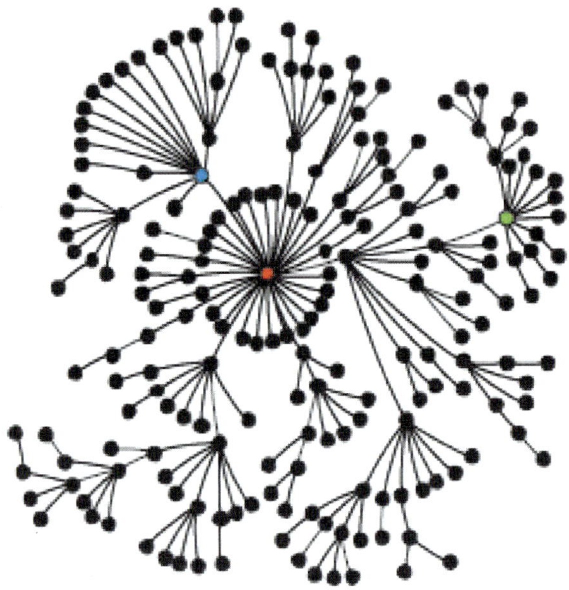

Fig. 2. A Scale-free large agent network [16]

the outbound links to generate the total number of links. When the total link number of all the agents is added, the number is duplicated because one link has two nodes and thus each link is counted twice. Therefore, it needs to be divided by 2, which is the total number of links.

The damping factor in the general ranking process is different from its description in the PageRank algorithm [8,9]. The damping factor in the general ranking process is based on the ratio of the total number of an agent's inbound links to the total number of links of all agents associated with the R-agent (including the R-agent). The damping factor is calculated as follows:

$$D = \frac{2C_r}{\sum_{i=1}^{m} G_i + C_i} \tag{2}$$

where D $(0 < D < 1)$ denotes the damping factor, C_r denotes the number of an R-agent's inbound links in an AAG. The damping factor is affected by an R-agent's total number of inbound links. The importance of the other agent decreases when an R-agent's total number of inbound links increases. The coefficient of a P-agent candidate's visit frequency (VF) to an R-agent is considered in calculating the general ranking scores. The VF provides information of how often an R-agent cooperates with its P-agent candidates previously. We consider it a crucial and request-independent factor. In other words, the VF factor is not affected by the requests sent by an R-agent and the VF information helps to determine whether an agent's capability is out of date. There are two principles we employ in this preliminary stage including: (i) only the P-agent candidates' inbound links from the R-agent are considered, (ii) only recent visits are considered. The coefficient of a P-agent candidate's VF is calculated as follows:

$$f(V_p) = 1 + (V_p / \sum_{i=1}^{n} V_i) \tag{3}$$

where $f(V_p)$ denotes the coefficient of recent visits from an R-agent to a nominated P-agent candidate; V_p denotes the total number of recent visits from an R-agent to its nominated P-agent candidate p; n denotes the total number of P-agent candidates that have inbound links from their R-agent; V_i denotes the total number of recent visits from an R-agent to all its P-agent candidates on an AAG (i from 1 to n).

The reason for Eq. 3 is: the coefficient of the VF is important to the general ranking score; however, its impact on the score should not be unlimited. Therefore, the adoption of the ratio of V_p to V_i limits the impact of the VF factor and the ratio is always between 0 and 1. In addition, the VF factor is positive such that if there are more visits from the R-agent to its P-agent candidate, then this P-agent's general ranking score should be higher. The coefficient is always greater than 1. For a P-agent candidate that only has outbound links to its R-agent, then $f(V_p) = 1$. The response quality factor provides information about the quality of the service provided by a P-agent to its R-agent previously. During the process of dealing with a request from an R-agent, a selected P-agent does not always provide a response to the R-agent. The record of the response quality of a P-agent candidate affects the agent's reliability, which affects the agent's ranking score. The response quality issue in the agent matching area has been raised in previous work such as [3]. The response quality is calculated as follows. Let:

$$T_{avg} = (\sum_{i=1}^{n} T_i)/n_i,$$

denote the average response time of all the P-agent candidates communicating with the R-agent previously. The response quality is given by,

$$Q_p = (T_{avg}/T_p) \times (0.5 + 1/2L_p) \tag{4}$$

where Q_p denotes P-agent candidate X's average response quality to an R-agent (it represents the P-agent's response quality to its R-agent); Ti denotes each individual P-agent candidate's average response time to its R-agent (these P-agent candidates have inbound links from their R-agents) and i is a variable; T_p denotes X's average response time to its R-agent (T_p is the average response time of all the requests from R-agent dealt with by X); L_p denotes the number of links that X established with other agents to solve a request from its R-agent. The reason for the inclusion of L_p is that: when X cannot resolve a request from an R-agent independently and demands other agents' involvements, then the reliability of the results decrease because additional agents are involved. For a

P-agent candidate that only has outbound links with its R-agent, which is the average value of the other agents' Lp in the same AAG. The reason for using $(1+1/L_p)/2$ is to restrain the value of Q_p.

3.2 Composition of General Ranking

Based on the previous equations, the general ranking equation can be deduced as follows:

$$GR(A) = \underbrace{\left((1-D) + D \times \left(\sum_j \frac{GR_j}{G_j} + GR_R\right)\right)}_{Part1} \times \underbrace{Q_A \times f(V_A)}_{Part2}$$

where $GR(A)$ denotes the general ranking score of a P-agent candidate A (A hereafter); GR_j denotes the general ranking score of P-agent candidate j that points its outbound link towards A (votes A); G_j denotes the outbound links of P-agent candidate j; j is a variable that denotes the P-agent candidates vote A; GR_{RA} denotes the ranking score of the R-agent that has outbound links to the P-agent candidates; QA is the response quality of P-agent A to its R-agent; $f(VA)$ is the coefficient of A's visit frequency. If the selected P-agent has no inbound links from its R-agent, then $GR_{RA} = 0$. Otherwise,

$$GR_{RA} = \sum_j \frac{GR_j}{n}.$$

where n denotes the total number of P-agent candidates; GR_{RA} denotes the general ranking score of an R-agent. R-agent's general ranking score is the average value of all its P-agent candidates' general ranking score. The reason to count R-agent's rank into its general ranking score is that: if a P-agent candidate receives inbound links from its R-agent then the general ranking score of this agent increases. Thus, R-agent has an impact on an agent's general ranking score. Note that the determination of $GR(A)$ is an iterative process, since GR_j in the equation will in turn depend on $GR(A)$. The general ranking equation can be viewed as two parts. Part 1 calculates A's importance (ranking score) among the other P-agent candidates and part 2 calculates A's ranking score based on its previous relationships with R-agent.

3.3 Example of General Ranking Calculation and Experimental Results

The general ranking process has been illustrated in [17]. The following example is provided based on Fig.1. The values of the P-agent candidates in Fig.1 are listed in Table1. V_i, T_i, and L_i are based on previous simulation values; G_i and C_i are based on Fig.1.

Table 1. Example values of P-agents in an AAG

P-agent	G_i	C_i	V_i	T_i (sec.)	L_i
A1	1	3	22	60	1
A2	2	1	8	102	3
A3	1	1	$-^*$	$-^\dagger$	$-^\dagger$
A4	1	2	2	396	4
A5	2	1	30	90	2
A6	0	3	11	50	1
A7	1	0	$-^*$	$-^\dagger$	$-^\dagger$
R-agent	5	2	–	–	–

$*f(V_{A3,7}) = 1$ (due to the inbound link from R-agent $= 0$)
$\dagger$ T_i = average of all the other non-null values. The column L_i gives the number of cooperating agents involved in solving a request for a selected P-agent; the values in this column are given and not explicitly indicated in Fig.1. For instance, $L_i = 3$ in A2 signifies that: A2 needs three other cooperating agents to solve the request from the R-agent. The general ranking score of the agent A1 in Fig.1 can be calculated as follows.

$$E = G_{A1,\ldots,7,R} + C_{A1,\ldots,7,R} = 13; D = C_{A1}/E = 3/13 = 0.23;$$

The coefficient of visit frequency can be calculated as follows:

$$f(V_{A1}) = 1 + (V_{A1}/(V_{A1} + V_{A2} + V_{A4} + V_{A5} + V_{A6})) \approx 1.3; = 1.11;$$

$$f(V_{A3}) = f(V_{A7}) = 1; f(V_{A4}) \approx 1.03; f(V_{A5}) = 1.411; f(V_{A6}) \approx 1.15;$$

The average response time can be calculated as:

$$T_{avg} = (T_{A1} + T_{A2} + T_{A4} + T_{A5} + T_{A6})/5 = 738/5 = 147.6;$$

Thus, the response qualities of the P-agent candidates are:

$$Q_{A1} = (T_{avg}/T_{A1}) \times (0.5 + 1/2L_{A1}) = 2.46; Q_{A2} \approx 0.97; Q_{A3} \approx 0.73;$$

$$Q_{A4} \approx 0.23; Q_{A5} = 1.23; Q_{A6} \approx 2.95; Q_{A7} \approx 0.73;$$

A system has been developed to calculate the general ranking scores of the above example iteratively. The initial values of all the P-agent candidates are 0. In this example, the preliminary iterations value is 50, which means all the P-agent candidates' general ranking scores are calculated 50 times in each general ranking process. After 50 iterations, all the general ranking scores become stable, see [17] for details.

Table 2. Final general ranking scores of the example

P-agent	General ranking score
A1	5.73
A2	1.45
A3	1.52
A4	0.36
A5	2.35
A6	5.80
A7	0.56

Table2 shows the final general ranking scores of the P-agent candidates based on the above example. A6 ranks highest in the table, followed closely by A1. The results indicate that the general ranking scores become stable after running 50 iterations. The ranking scores stabilise rather quickly after 10 iterations, although some agents (A4 and A7) reach the stable values much sooner.

After 50 iterations, all the general ranking scores become stable. Table2 shows the final general ranking scores of the P-agent candidates based on the above example.

We see that A6 ranks highest, followed closely by A1. There is a substantial gap between A1 and the next one A5, while the lowest ranking agents are A7 and A4. The ranking scores stabilise rather quickly after 10 iterations, although some agents (A4 and A7) reach the stable values much sooner.

The general ranking process can expand to some more complex AAGs such as a quasigroup form of AAG or other complex network graphs with a central node (R-agent). In spite of varied forms of AAGs, the calculation process in the general ranking process remains the same. In addition, a two-way link (with inbound and outbound links) between two agents in an AAG is regarded as two links, which include one inbound and one outbound link.

4 The AR Algorithm-Part II: Request-Based Ranking

The general ranking scores are based on the history of P-agent candidates' services. Request-based ranking (referred to as RBR hereafter) disregards previous relationships between R-agent and P-agent candidates and it only calculates the ranking scores based on a particular request.

In [18], a Weighted-Tree Similarity (WTS) algorithm is introduced for agent matching in the E-business environment. Despite the limitations of the WTS algorithm, the basic idea behind the WTS seems useful in the agent matching process. The RBR calculation process adopts the similarity tree concept, in which a request is decomposed into several key terms and each key term is further extended to be a more specific and comprehensive description. Through the semantic extension process, a request from R-agent turns into a semantic extension tree and the RBR scores are based on the calculation of the end nodes.

A request is decomposed through a reasoning process and forms a semantic extension tree. Some extensions can be further expanded and added to the tree. The final semantic extension tree is completed when it exhausts all the extensions of each node.

The process of constructing a RBR semantic extension tree or an agent capability description tree is based on existing semantic networks [19] and ontology technology. There are several reasons for adopting a tree-based similarity calculation method in agent matching in preference to using existing text mining or Web-based semantic matching technology, and they are listed as follows:

First, a large amount of time delay resulting from an unsuccessful agent matching process necessitates high matching accuracy. The time delay in an unsuccessful agent matching process is much greater than in an unsuccessful Web page search. This is because a false match is only detected in an agent matching process when a P-agent fails to handle the request from its R-agent. For instance, a P-agent spends a great deal of time dealing with a request but finally fails to provide some useful results. Therefore, conducting a pre-evaluation based on an agent's overall capabilities to find the most (or nearly the most) appropriate agent is necessary. The time delay for a re-match in the Web-based searching process is much lower than that of agent matching. Thus, efficient similarity calculation in agent matching is crucial.

Second, current agent description languages cannot distinguish agent capabilities efficiently as many agents, especially in a large multi-agent system, seem to provide very similar descriptions. Therefore, an agent matching process requires more accurate and detailed analysis compared to simple text mining.

Finally, the tree-like descriptions are efficient representations of an agent's capabilities. An agent's capabilities and functions can be efficiently broken down into more specific details by using a tree-like representation method.

4.1 Agent Capability Description Tree

In [4], an agent description language, LARKS, is deployed to describe an agent's capability and to provide an efficient agent service matching mechanism. In [20], agent capabilities are formalised in a Beliefs-Desire-Intentions (BDI) framework. Based on previous work, we suggest a top-down folding tree to classify agent capabilities.

In a top-down folding tree, the root of the tree describes an agent's capability set and each successive level parent node represents the capability of an agent, where parent nodes represent the sub-domains to which the child nodes belong. The leaf nodes define the general categories to which the agent's capability belongs. The middle level nodes and leaf nodes in a description tree can be folded, which enhances the efficiency in describing agent's capabilities by omitting irrelevant or unimportant information. The category classification of the fields of knowledge of science, social science and arts, and humanities adopts various standards such as [21].

4.2 RBR Similarity Calculation

An extraction process is deployed to collect two information sets from two trees while calculating the similarity between a RBR semantic extension tree and an agent capability description tree. Each set contains a number of semantic elements and coefficients representing the node information in both trees. The similarity calculation is based on semantic matching between the two sets and coefficient calculation of the matched elements. The RBR similarity calculation process makes use of some concepts from existing data mining and information retrieval technologies include keyword-based retrieval, similarity-based retrieval and classification analysis [22].

The *Cosine-Similarity Measure* (CSM) [23,24] has been applied extensively in the information retrieval domain to calculate the similarity between two vectors, which is their inner product normalised by their lengths,

$$sim_{cos}(X,Y) = \frac{X \cdot Y}{\mid X \mid\mid Y \mid} \tag{5}$$

However, the CSM cannot function efficiently when the dimensions of these vectors are unequal. To overcome this, we adopt the *Cartesian-Product* [25] method in the RBR-based similarity calculation between two sets, which is defined as:

$$A \times B = \{(a,b) \mid a \in A \; and \; b \in B\} \tag{6}$$

where A contains the elements extracted from a RBR semantic tree; B contains the elements extracted from an agent capability description tree. The RBR-based similarity calculation consists of three main steps. The first step generates a Cartesian product; the second step calculates the similarity value of two elements in the Cartesian product; the final step sums the level coefficie and the path coefficient of a pair of matched elements through incorporating their similarity values. For instance, consider two sets S_1 and S_2, which are extracted from a RBR semantic extension tree (S_1) and an agent capability description tree (S_2), respectively.

The first step is to generate the Cartesian product by using the agent capability description tree and a RBR tree. Let:

$$S_1 = \{l_1 A_1, l_2 A_2, \ldots l_i A_i\}, S_2 = \left\{\frac{1}{P_1} B_1, \frac{1}{P_2} B_2, \ldots \frac{1}{P_j} B_j\right\},$$

where A_i is an element in an agent capability description tree; B_j is an element in the RBR extension tree; li denotes the level coefficient of A_i; P_j denotes path coefficient of B_j and N denotes the total extension number of a nominated node's parent node. For a RBR extension tree, the similarity value decreases when the level number increases, whereas the similarity value increases when the path number increases in an agent capability description tree. Then, refer to the root node as level 0 (or path 0), its child nodes as level 1, and the rest of the nodes' level number is increased by 1 from their parents' level number. Let $i = 3$, $j = 4$, thus the Cartesian-product of set S_1 and S_2 is:

$$S_1 \times S_2 = \Big\{(l_1A_1, \frac{1}{P_1}B_1), (l_1A_1, \frac{1}{P_2}B_2), (l_1A_1, \frac{1}{P_3}B_3), (l_1A_1, \frac{1}{P_4}B_4),$$
$$(l_2A_2, \frac{1}{P_1}B_1), (l_2A_2, \frac{1}{P_2}B_2), (l_2A_2, \frac{1}{P_3}B_3), (l_2A_2, \frac{1}{P_4}B_4),$$
$$(l_3A_3, \frac{1}{P_1}B_1), (l_3A_3, \frac{1}{P_2}B_2), (l_3A_3, \frac{1}{P_3}B_3), (l_2A_2, \frac{1}{P_4}B_4)\Big\}$$

The next step is to calculate the similarity value of two elements (words) in a Cartesian product. In the RBR-based similarity calculation process, every agent description tree is fully expanded and every node is calculated in the Cartesian product. The methods used in constructing descriptive trees and comparing elements between semantic elements can be diverse according to existing semantic technology. Several projects have been carried out in this area, such as WordNet [27]. Here, a three-type semantic relationship model based on WordNet is deployed to describe the similarity relationships between two words. The three types of relationships [27,28] are: (i) Synonym: Two words are synonymous. (ii) IS-a: Two words are in a superset and subset relationship. (iii) Has-a: One word has ownership of another word. Also known as part-whole relationship between words.

In this example, if two words are synonyms, then the similarity value is 1; if two words have an 'IS-a' relationship then the similarity value is 0.9; if two words have a 'Has-a' relationship then the similarity value is 0.85. The initialisation value of each relationship type is based on previous experience, and these values can be adjusted according to different applications. If two words match, their upper (or lower in the case of agent capability description tree) level words are required for further matching. The comparison process continues until the entire upper or lower levels are matched.

An algorithm (Algorithm-1) is provided below to demonstrate the process of calculating the similarity values in other levels related to the two matched words. In this algorithm, S denotes the similarity value of a Cartesian product (two elements); $NodeX$ denotes a node in the RBR extension tree; $NodeY$ denotes a node in an agent capability description tree; $Synonym(x, y)$ is the function that determines whether two words are synonyms; $ISA(x, y)$ is the function determining whether two words are hypernyms/hyponyms; $HAS(x, y)$ is the function determining whether two words are meronyms/holonyms; $NodeSelection(x)$ is the function that determines which node is the next level node to move to. $MoveUpper(x)$ is the function that the selected node moves to the upper level node in the tree for matching; $MoveLower(y)$ is the function that the selected node moves to the lower level node in the tree for matching. If two words do not match then exit the process (if at the starting point then $S = 0$; else S remains the last calculated value). If there are more than two upper or lower level nodes, then select the highest similarity value.

Algorithm -1: Similarity Value Calculation

```
S := 1; /* lower level match for two nodes */
while(NodeX ≠ Root)and(NodeY ≠ Leaf)do
    if Synonym(NodeX, NodeY) := true then
        {S = (S + 1)/2;}
    else if ISA(NodeX, NodeY) := true then
        {S = (S + 0.9)/2;}
    else if HAS(NodeX, NodeY) := true then
        {S = (S + 0.85)/2;}
    else
        {S = (S + 0.1)/2;}
    /*if other level nodes not match S = 0.1*/
    Node(X) := NodeSelection(NodeX + 1);
    Node(Y) := NodeSelection(NodeY − 1);
    MoveUpper(NodeX);
    MoveLower(NodeY);
    return(S);
end.
```

The final step is to extract level and path coefficients and performing an equivalence operation between two elements in a Cartesian product to calculate the RBR-based similarity value. The RBR-based similarity score (value) is calculated as:

$$RS = \frac{\left(\sum_{i \in R, j \in C} \frac{l_i}{P_j} \times S_{i,j}\right)}{M} = \frac{\left(\sum_{i \in R, j \in C} \frac{l_i}{P_j} \times (S_d + S_x)\right)}{M} \tag{7}$$

Here, RS denotes the final RBR-based similarity score; C denotes total level number of a RBR semantic extension tree, R denotes the total level number of an agent capability description tree; $S_{i,j}$ denotes the composite similarity value; M denotes the total element number in the Cartesian product; S_d denotes the independent similarity value of a Cartesian product without considering other level nodes; S_x denotes the total similarity values of other nodes that are ancestors or descendants of the two nodes in the Cartesian product; l_i denotes the level number of a node in an agent capability tree; P_i denotes the level number of a node in a RBR semantic tree. For instance, supposing $Synonym(A_1, B_1) = true$; $HAS(A_1, B_4) = true$; $Synonym(A_2, B_2) = true$; $ISA(A_2, B_3) = true$; $ISA(A_3, B_1) = true$; $Synonym(A_3, B_3) = true$. We have the results shown in Table3.

In Table3, *Independent Similarity Value* indicates the similarity calculation of a Cartesian product without considering the nodes in other levels. *Composite Similarity Value* indicates the final similarity value of a Cartesian product that incorporates the similarity value of the nodes in other levels. Thus, we have the similarity values for all the Cartesian products:

Table 3. RBR Similarity Value (SV) for Set A and B

Cartesian Product	Independent SV	Composite SV
A_1, B_1	1	1
A_1, B_2	0	0
A_1, B_3	0	0
A_1, B_4	0.85	0.85
A_2, B_1	0	0
A_2, B_2	1	1
A_2, B_3	0.9	0.875
A_2, B_4	0	0
A_3, B_1	0.9	0.9
A_3, B_2	0	0
A_3, B_3	1	0.925
A_2, B_4	0	0

$$S_1 \times S_2 = \left\{ \begin{array}{cccc} (\frac{l_1}{P_1}), & 0, & 0, & 0.85(\frac{l_1}{P_4}), \\ 0, & (\frac{l_2}{P_2}), & 0.875(\frac{l_3}{P_3}), & 0, \\ 0.9(\frac{l_3}{P_1}), & 0, & 0.925(\frac{l_3}{P_3}), & 0 \end{array} \right\}$$

Therefore, the final similarity score of two trees is:

$$RS = \left\{ (\frac{l_1}{P_1}) + 0.85(\frac{l_1}{P_4}) + (\frac{l_2}{P_2}) + 0.875(\frac{l_3}{P_3}) + 0.9(\frac{l_3}{P_1}) + 0.925(\frac{l_3}{P_3}) \right\} /12$$

Overall, the RBR-based similarity calculation process follows the two main equations (6) and (7). Various agents possess different agent capability description trees, and the RBR process ranks the similarity scores of all the agent capability description trees. Fig.3 and Fig.4 show the RBR calculation scores based on the experiments conducted in [17] and [28].

The final agent ranking scores combine the general ranking scores and the RBR similarity scores. In general, the overall agent-ranking score can be calculated using the convex combination:

$$AR = \alpha \times GR + (1 - \alpha) \times RS \tag{8}$$

where α $(0 < \alpha < 1)$ is the weight assigned to GR, which denotes the general ranking score; RS denotes request-based ranking score which receives a weight of $(1 - \alpha)$; and the AR gives the overall agent-ranking score. The final score α could be set as a specific ratio, such as 0.6, to indicate how much weight that GR should be assigned GR and RS. After the calculation process, A1 is the most appropriate P-agent to R-agent RA among the four trees, although it has second highest score in general ranking. A6 has highest score in general ranking but is less relevant to RA in request-based ranking. Thus, A1 is selected as the P-agent.

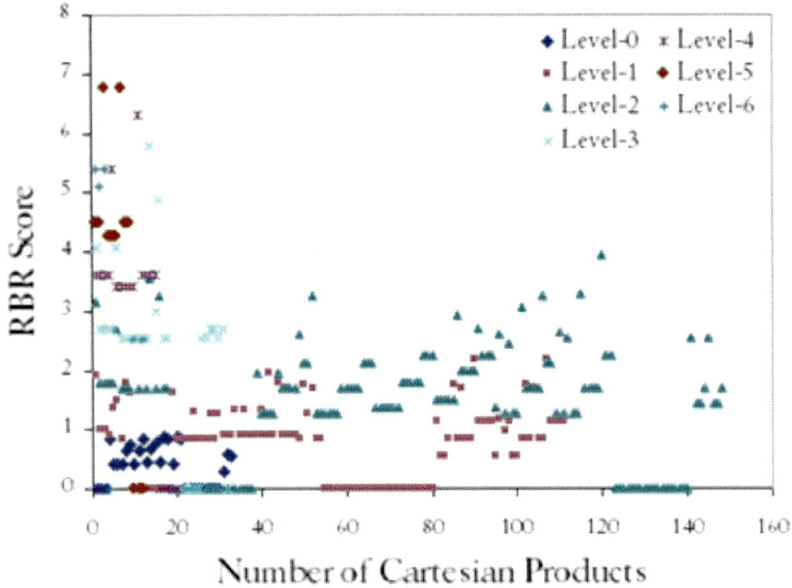

Fig. 3. The RBR Similarity Score of A1

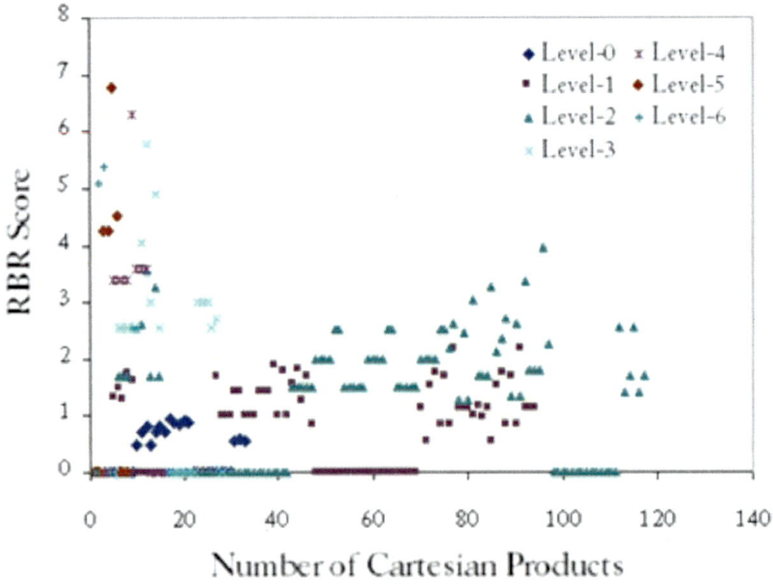

Fig. 4. The RBR Similarity Score of A6

5 Conclusion

Searching a proper service for a particular request is a common and essential process in modern life. In many service-based searching processes, some irregular services and their potential relationships with other services and requests are often overlooked. This issue has been raised in Chance Discovery theory [29], which leads to the further investigation on how to efficiently discover known and unknown services and associate their potential relationships.

This research investigates service-matching processes based on multi-agent systems and introduces the AR algorithm to enhance the matching efficiency. This algorithm avoids massive searching in a complex system, which is particularly efficient in a versatile large scale multi-agent system. In such a system, the effective matching and deployment of agents will have a critical impact on overall problem solving performance. The agent-matching problem has a different level of complexity compared with the searching and matching of web pages. Over and above the linear matching of strings to determine relevance, the discovery of functions and capabilities is required. Here, we use a methodology whereby past relationships and performance are taken into account.

The AR algorithm provides an efficient methodology to solve the problem that plague the current Chance Discovery processes [30,31] through collecting essential participants (R-agents) and match them with corresponding service-providers (i.e. P-agents). The AR algorithm consists of two major processes: the general ranking process and request-based ranking process. The matching process is based on similarity computations on the semantic trees. The P-agent candidate with highest similarity score is more likely to be selected as the P-agent.

The AR algorithm improves the success rate of searching proper services for a request. Although the present algorithm does carry some ranking computation overhead, such computation cost is considered acceptable in comparison with the penalty of inappropriate matching which may result in defective solutions and time wasted in carrying out re-matching. The future work of this research can be extended to apply the AR algorithm to discover the hidden relationships among various services.

Acknowledgement. This research work is supported by the Zhejiang Philosophy and Social Science Project Fund (Grant No. 11JCSH03YB) and the National Natural Science Fund of China (Grant No. 70871111).

References

1. Biskupski, B., Dowling, J., Sacha, J.: Properties and Mechanisms of Self-Organizing MANET and P2P Systems. ACM Transactions on Autonomous and Adaptive Systems 2(1), 1–34 (2007)
2. Parunak, H.V.D., Brueckner, S.A., Matthews, R.: Pheromone Learning for Self-Organizing Agents. IEEE Transactions on SMC-A 35(3), 316–326 (2005)
3. Li, X., Montazemi, A.R., Yuan, Y.: Agent-based buddy-finding methodology for knowledge sharing. Information & Management 43, 283–296 (2006)

4. Sycara, K., Widoff, S.: LARKS: Dynamic Matchmaking Among Heterogeneous Software Agents in Cyberspace. In: Autonomous Agents and Multi-Agent Systems, vol. 5, pp. 173–203. Kluwer Academic Publishers (2002)
5. Sycara, K., Paolucci, M., Ankolekar, A., Srinivasan, N.: Automated Discovery, Interaction and Composition of Semantic Web services. Journal of Web Semantics 1(1), 27–46 (2003)
6. Shvaiko, P., Euzenat, J.: A survey of schema-based matching approaches. Journal on Data Semantics 4, 146–171 (2005)
7. Halevy, A.: Answering queries using views: a survey. VLDB Journal 10(4), 270–294 (2001)
8. Farahat, A., LoFaro, T., Miller, J.C., Rae, G., Ward, L.A.: Authority Rankings from HITS, PageRank, and SALSA: Existence, Uniqueness, and Effect of Initialization? SIAM Journal on Scientific Computing 27(4), 1181–1201 (2006)
9. Page, L., Brin, S., Motwani, R., Winograd, T.: The PageRank citation ranking: Bringing order to the web, Technical report, Stanford University, Stanford, CA (1998)
10. Zhang, H.L., Leung, C.H.C., Raikundalia, G.K.: Topological analysis of AOCD-based agent networks and experimental results. Journal of Computer and System Sciences 74, 255–278 (2008)
11. Zhang, H.L., Leung, C.H.C., Raikundalia, G.K.: Matrix-Agent Framework: A Virtual Platform for Multi-agents. Journal of System Sciences and Systems Engineering 15(4), 436–456 (2006)
12. Sarne, D., Kraus, S.: Time-Variant Distributed Agent Matching Applications. In: Proc. of AAMAS, vol. 1, pp. 168–175. IEEE CS Press (2004)
13. Kebriaei, H., Majd, V.J., Rahimi-Kian, A.: A New Agent Scheme Using an Ordered Fuzzy Similarity Measure And Game Theory. Computational Intelligence 24(2), 108–121 (2008)
14. Subrahmanian, V., Bonatti, P., et al.: Heterogeneous Agent Systems, pp. 43–59 (2000)
15. Kuhn, N.: Comparing Ranking of Heterogeneous Agentsm. In: Proc. of COOCS, pp. 1–12. ACM Press (1993)
16. Strogatz, S.H.: Exploring complex networks. Nature 410, 440–442 (2001)
17. Zhang, H.L., Leung, C.H.C., Raikundalia, G.K., He, J.: A Novel Ranking Algorithm for Service Matching Based on Agent Association Graphs. In: Proc. of IEEE ICDM Workshop, pp. 1273–1280. IEEE Press (2010)
18. Bhavsar, V.C., Boley, H., Yang, L.: A Weighted-Tree Similarity Algorithm for Multi-agent Systems in E-business Environments. Computational Intelligence 20(4), 584–602 (2004)
19. Sowa, J.F.: Principles of Semantic Networks? Explorations in the representation of knowledge. Morgan Kaufman Publishers (1991)
20. Padgham, L., Lambrix, P.: Agent Capabilities: Extending BDI Theory. In: Proc. of the 17th AAAI, pp. 68–73. AAAI Press & The MIT Press (2000)
21. Glanzel, W., Schubert, A.: A new classification scheme of science fields and subfields design for scientometric evaluation purpose. Scientometrics 56(3), 357–367 (2003)
22. Han, J., Kamber, M., Pei, J.: Data Mining: Concept and Techniques, 3rd edn. Morgan Kaufman Publishers (2011)
23. Ganesan, P., Garcia-Molina, H., Widom, J.: Exploiting Hierarchical Domain Structure to Compute Similarity. ACM Transactions on Information Systems 21(1), 64–93 (2003)
24. Salton, G., Buckley, C.: Term-weighting approaches in automatic text retrieval. Inf. Process. Manage. 24(5), 513–523 (1988)

25. Hein, J.L.: Discrete Mathematics, pp. 74–140. Jones and Bartlett Publishers (2002)
26. Fellbaum, C. (ed.): WordNet: A Lexical Database for English. MIT Press (1998)
27. Li, Y., Bandar, Z.A., McLean, D.: An approach for measuring semantic similarity between words using multiple information sources. IEEE Transaction on Knowledge and Data Engineering 15(4), 871–882 (2003)
28. Zhang, H.L., Leung, C.H.C., Tang, X.: Discovering and Matching Service Providers among Heterogeneous Agents. In: Proc. of IJCAI Workshop, pp. 27–32 (2011)
29. Abe, A., Ohsawa, Y. (eds.): Readings in Chance Discovery. International Series on Advanced Intelligence (2005)
30. McBurney, P., Ohsawa, Y. (eds.): Chance Discovery. Advanced Information Processing (2003), ISBN 3-540-00549-8
31. Ohsawa, Y., Nishihara, Y.: Innovators' Marketplace: Using Games to Activate and Train Innovators. Springer (forthcoming, 2012)

Discovering the Chances of Health Problems and Falls in the Elderly Using Data Mining

Bogdan Pogorelc and Matjaž Gams

Jožef Stefan Institute, Ljubljana, Slovenia
Špica International d.o.o., Ljubljana, Slovenia
bogdan.pogorelc, matjaz.gams@ijs.si

Abstract. The basis for this work was an invited paper at the International Joint Conference on Artificial Intelligence (IJCAI) workshop on Chance Discovery in 2011 and an additional invitation to submit an extended version of a paper to this book. Here we present a generalized approach to the detection of the chances of health problems and falls in the elderly for the purpose of prolonging their autonomous living using a novel data-mining approach. The movement of the user is captured with a motion-capture system that consists of body-worn tags, whose coordinates are acquired by sensors located in an apartment. The output time series of the coordinates are modeled with the proposed data-mining approach in order to recognize the specific health problem or fall. The approach is general in the sense that it uses a k-nearest-neighbor algorithm and dynamic time warping with the time series of all the measurable joint angles for the attributes instead of a more specific approach with medically defined attributes. It is a two-step approach: in the first step it classifies the person's activities into five activities, including different types of falls. In the second step it classifies classified walking instances from the first step into five different health states: one healthy and four unhealthy. Even though the new approach is more general and can be used to differentiate other types of activities or health problems, it achieves very high classification accuracies, similar to the more specific approaches described in the literature.

Keywords: health problems, activities, falls, elderly, machine learning, data mining.

1 Introduction

The number of elderly people in the developed countries is increasing [20], and they tend to lead isolated lives away from their offspring. In many cases they fear being unable to obtain help if they are injured or ill.

In recent decades this fear has resulted in research attempts to find assistive technologies to make the living of elderly people easier and more independent. The aim of this study is to provide ambient assistive-living services to improve the quality of life of older adults living at home.

Y. Ohsawa & A. Abe (Eds.): Advances in Chance Discovery, SCI 423, pp. 163–175.
springerlink.com

We propose a generalized approach to an intelligent and ubiquitous care system to recognize a few of the most common and important health problems in the elderly, which can be detected by observing and analyzing the characteristics of their movement.

It is two-step approach as shown in Fig. 1. In the first step it classifies the person's activities into five activities, including two types of falls. In the second step it classifies classified walking instances from the first step into five different health states: one healthy (N) and four unhealthy.
The activities are:

- fall (F),
- unconscious fall (UF),
- walking (W),
- standing/sitting (SS),
- lying down/lying (L).

The types of abnormal health states are:

- hemiplegia (usually the result of a stroke, H),
- Parkinson's disease (P),
- pain in the leg (L),
- pain in the back (B).

The movement of the user is captured with a motion-capture system, which consists of tags attached to the body, whose coordinates are acquired by sensors located in the apartment. The output time series of the coordinates are modeled with the proposed data-mining approach in order to recognize the specific activity or health problem. The architecture of the system is presented in Fig. 1.

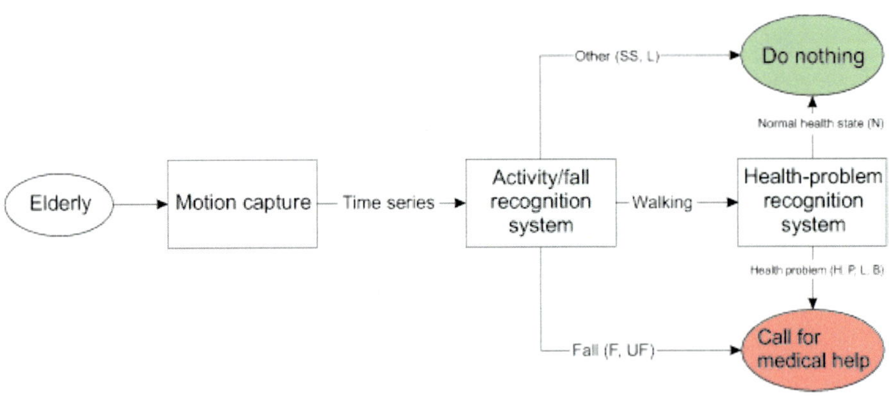

Fig. 1. Architecture of the system

1.1 Relevance of the Presented Work for Chance Discovery

A Chance Discovery is the discovery of chance. In the presented study it is the discovery of a chance that an elderly person has fallen, that he/she has a health problem or that he/she is performing one of the activities of daily living (ADL).

Chance in the Chance Discovery research field is defined as a new event/situation that can be conceived either as an opportunity or a risk for the future. Risks for the future that the presented system is discovering or reducing are:

- the risk that a person has a health problem,
- the risk that the health problem progresses, from worsening the quality and independency of living to a death,
- the risk that a person has fallen, which can also worsen the quality of living or cause death,
- the risk that a person does not receive treatment quickly enough.

It is also making possible certain opportunities:

- the opportunity that an elderly person rehabilitates better because of receiving the diagnosis and, consequently, the treatment on time,
- the opportunity that an elderly person has more secure and independent living at his/her home.

Moreover, the preventive measures are taken in the case of a discovered chance of a health problem or fall through the notification of a medical center.

Chance discovery is research that studies how to discover rare events, causing a potentially significant situation. These rare events are the abnormal gait patterns that the system is detecting and are potentially related to health problems. The inductive chance-discovery technique, i.e., data mining, is the core of the presented system.

1.2 Organization of the Study

The presented study is organized as follows. The second section provides a review of the related work in the field. The third section gives a description of the data and the methods that were used in the study. The fourth section presents the experiments and results, supported by tables. The last section concludes the study and provides ideas for further work.

2 Related Work

In related studies the motion is normally captured with inertial sensors [19,1], computer vision and also with a specific sensor for measuring the angle of joint deflection [15] or with electromyography [21]. In our study an infra-red (IR) sensor system with tags attached to the body [5] was used.

We do not only address the recognition of activities of daily living, such as walking, sitting, lying, etc. and the detection of falling, which has been addressed many times [3,10], but also the recognition of health problems based on motion data.

Using a similar motion-capture system to that in our approach, the automatic distinction between health problems such as hemiplegia and diplegia is presented [9]. However, a much more common approach to the recognition of health problems is the capturing of movement that is later manually examined by medical experts [15,4,12]. Such an approach has a major drawback in comparison to ours, because it needs to be constantly monitored by medical professionals.

The paper [11] presented a review of assistive technologies for care of the elderly. The first technology consists of a set of alarm systems installed at people's homes. The system includes a device in the form of a mobile phone, a pendant or a chainlet that has an alarm button. They are used to alert and communicate with a warden. When the warden is not available, the alert is sent to the control center. However, such devices are efficient only if the person recognizes the emergency and has the physical and mental capacity to press the alarm button.

The second technology presented in [11] is video-monitoring. The audio-video communication is done in real time over an ordinary telephone line. The video can be viewed on a monitor or domestic television. The problems of the presented solution are ethical issues, since elderly users do not want to be monitored by video [3]. Moreover, such an approach requires the constant attention of the emergency center.

Miskelly [11] also presented a technology based on health monitors. The health monitor is worn on the wrist and continuously monitors the pulse, skin temperature and movement. At the beginning of the system's use, the pattern for the user is learned. Afterwards, any deviations are detected and alarms are sent to the emergency center. Such a system detects collapses, faints, blackouts, etc.

Another presented technology is the group of fall detectors. They measure the accelerations of the person using tags worn around the waist or the upper chest. If the accelerations exceed a threshold during a time period, an alarm is raised and sent to the community alarm service. Bourke et al. [2] presented the acceleration data produced during the activities of daily living and when a person falls. The data was acquired by monitoring young subjects performing simulated falls. In addition, elderly people performed the activities of daily living. Then, by defining the appropriate threshold it is possible to distinguish between the accelerations during falls and the accelerations produced during the normal activities of daily living. In this way accelerometers with a threshold can be used to monitor elderly people and recognize falls. However, threshold-based algorithms produce mistakes, for instance, quickly standing up from or sitting down on a chair could result in crossing the threshold, which is erroneously recognized as a fall.

Rudel [16] proposed the architecture of a system that enables the control of users in their homes. It consists of three levels. The first level represents the ill people in their homes equipped with communication and measurement devices. The second level is the information and communication technology that enables the communication with the main server. The third level is the telemedicine center, including the duty operator, doctors and technical support, the center for the implementation of direct assistance at home, and the team of experts for implementing the telemedicine services. Such a system does not provide any automatic detection of unusual behavior but instead requires constant observation by the medical center.

Williams et al. [22] have shown that the ability to perform daily activities is reduced for people that have fallen several times and that this reduction can be detected using accelerometers. They tested elderly people that had not fallen and those that had fallen several times. All of them were asked to perform a predefined scenario, including sentence writing, object picking, etc. The accelerations differ significantly between the two groups of people during the test.

Perolle et al. [13] described an elderly-care system that consists of a mobile module worn by the user all the time that is able to locate the user, detect falls and monitor the user's activity. In addition, this device is connected to a call center, where the data is collected, analyzed, and emergency situations are managed. The mobile module is worn on a belt. It produces an alarm, provides the possibility to cancel it, shows the battery status, etc. In addition, it monitors the user activity and gives it three classifications: low, medium and high. Once a day, the data is sent to the call center for analysis. The user is located with a GPS, for when it is necessary to respond to alarms and to locate the user if he/she gets lost. The mobile module also provides bidirectional voice communication between the user and the call center in order to communicate critical information immediately.

The studies [14,23] differentiate between the same five health states as presented in this study, but are more specific due to the use of 13 medically defined attributes. The currently presented study instead uses very general attributes of the angles between body parts, allowing the system to use the same attributes and the same classification methods for differentiating between five activities and between five health states.

The aim of this study is to realize an automatic classifier that is able to support the autonomous living of the elderly by detecting falls and health problems that are recognizable through movement. Earlier works (e.g., [7]) describe machine-learning techniques employed to analyze activities based on the static positions and recognized postures of the users. Although these kinds of approaches can leverage a wealth of machine-learning techniques, they fail to take into account the dynamics of the movement.

The present work has instead the aim to recognize movements by observing the time series of the movements of the users. Better activity-recognition performance can be achieved by using pattern-matching techniques, which take into account all of the sensors' readings, in parallel, considering their time course.

3 Materials and Methods

3.1 Targeted Activities and Health Problems for Detection

The proposed system uses a two-step approach for the recognition of important situations. All the situations that we are recognizing were suggested by the collaborating medical expert on the basis of occurrence in the elderly aged over 65, the medical significance and the feasibility of their recognition from movements. Thus, in the first step we are recognizing five activities: accidental fall, unconscious fall, walking, standing/sitting, lying down/lying. We are focusing on differentiating between "accidental fall" and "unconscious fall":

- **Accidental fall:** as the name suggests it happens due to an accident. The types of accidental falls are, e.g., stumbling and slipping. If the person does not hurt him/herself after it, he/she does not need medical attention.
- **Unconscious fall:** this happens due to an illness or a short loss of consciousness. In most cases the person who falls in this way needs medical attention.

The other three activities of interest are common activities at home, also known as the activities of daily living (ADL). In the second step we focused on four health problems and normal walking as a reference in accordance with the suggestions received from the collaborating medical expert. The following four health problems were chosen as the most appropriate [4]:

- **Parkinson's disease:** a degenerative disease of the brain (central nervous system) that often impairs motor skills, speech, and other functions. The symptoms are frequently tremor, rigidity and postural instability. The rate of the tremor is approximately 4–6 Hz. The tremor is present when the involved part(s), usually the arms or neck, are at rest. It is absent or diminished with sleep, sedation, and when performing skilled acts.
- **Hemiplegia:** is the paralysis of the arm, leg and torso on the same side of the body. It is usually the result of a stroke, although diseases affecting the spinal cord and the brain are also capable of producing this state. The paralysis hampers movement, especially walking, and can thus cause falls.
- **Pain in the leg:** resembles hemiplegia in that the step with one leg is different from the step with the other. In the elderly this usually means pain in the hip or in the knee.
- **Pain in the back:** this is similar to hemiplegia and pain in the leg in terms of the inequality of steps; however, the inequality is not as pronounced as in walking with pain in the leg.

The classification into five activities and into five health problems was made using the k-nearest-neighbor machine-learning algorithm and dynamic time warping for the similarity measure.

3.2 Attributes for Data Mining

The recordings consisted of the position coordinates for the 12 tags that were worn on the shoulders, the elbows, the wrists, the hips, the knees and the ankles, sampled at 10 Hz. The tag coordinates were acquired with a Smart IR motion-capture system with a 0.5-mm standard deviation of noise.

From the motion-capture system we obtain the position of each tag in x-y-z coordinates. Achieving the appropriate representation of the user's behavior activity was a challenging part of our research. The behavior needs to be represented by simple and general attributes, so that the classifier using these attributes will also be general and work well on behaviors that are different from those in our recordings. It is not difficult to design attributes specific to our recordings; such attributes would work well on them. However, since our recordings captured only a small part of the whole range of human behavior, overly specific attributes would likely fail on general behavior.

Considering the above mentioned, we designed attributes such as the angles between adjacent body parts. The angles between body parts that rotate in more than one direction are expressed with quaternions:

- q^t_{SL} and q^t_{SR}... left and right shoulder angles with respect to the upper torso at time t
- q^t_{HL} and q^t_{HR}... left and right hip angles with respect to the lower torso
- q^t_{TU} and q^t_{TL}... the angle (orientation) of the upper and of the lower torso
- α^t_{EL}, α^t_{ER}, α^t_{KL} and α^t_{KR}... left and right elbow angles, left and right knee angles.

3.3 Dynamic Time Warping

We will present dynamic time warping (DTW) as a robust technique to measure the "distance" between two time series [8]. Dynamic Time Warping aligns two time series in such a way that some distance measure is minimized (usually the Euclidean distance is used). Optimal alignment (minimum distance warp path) is obtained by allowing the assignment of multiple successive values of one time series to a single value of the other time series and therefore the DTW can also be calculated on time series of different lengths. Fig. 2 shows examples of two time series and the value alignment between them for the Euclidean distance (left) and the DTW similarity measure (right).

The time series have similar shapes, but are not aligned in time. While the Euclidean distance measure does not align the time series, the DTW does address the problem of time difference. By using DTW an optimal alignment is found among several different warp paths. This can be easily represented if two time series $A = (a_1, a_2, \ldots, a_n)$ and $B = (b_1, b_2, \ldots, b_m)$, $a_j, b_j \in R$ are arranged to form a n-by-m grid. Each grid point corresponds to an alignment between the elements $a_j \in A$ and $b_j \in B$. A warp path $W = w_1, w_2, \ldots, w_k, \ldots, w_K$ is a sequence of grid points where each w_k corresponds to a point $(i, j)_k$ — the warp path W maps elements of sequences A and B.

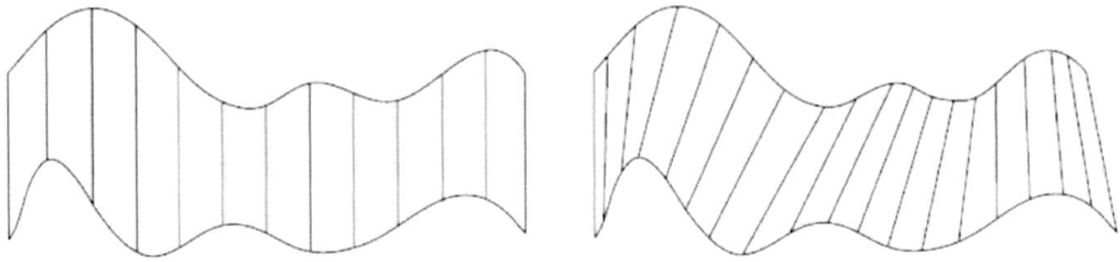

Fig. 2. Comparison of two similarity measures between two time series. The lines between the time series show the value alignment used by the Euclidean distance (left) and the Dynamic Time Warping similarity measure (right).

From all possible warp paths the DTW finds the optimal one [24]:

$$DTW(A,B) = min\ W\left[\sum_{k=1}^{K} d(w_k)\right]$$

The $d(w_k)$ is the distance between the elements of the time series.

The purpose of DTW is to find the minimum distance warp path between two time series. Dynamic programming can be used for this task. Instead of solving the entire problem all at once, solutions to sub-problems (sub-series) are found and used to repeatedly find the solution to a slightly larger problem. Let $DTW(A,B)$ be the distance of the optimal warp path between time series $A = (a_1, a_2, \ldots, a_n)$ and $B = (b_1, b_2, \ldots, b_m)$ and let $D(i,j) = DTW(A', B')$ be the distance of the optimal warp path between the prefixes of the time series A and B:

$$D(0,0) = 0$$

$$A' = (a_1, a_2, \ldots, a_i), B' = (b_1, b_2, \ldots, b_j)$$

$$0 \leq i \leq n, 0 \leq j \leq m$$

$DTW(A,B)$ can be calculated using the following recursive equations:

$$D(0,0) = 0$$

$$D(i,j) = min(D(i-1,j), D(j,j-1), D(i-1,j-1)) + d(a_i, b_j)$$

The distance between two values of the two time series (e.g. the Euclidean distance) is $d(a_i, b_j)$. The most common way of calculating $DTW(A,B)$ is to construct a $n*m$ cost matrix M, where each cell corresponds to the distance of the minimum distance warp path between the prefixes of the time series A and B (Fig. 3):

$$M(i,j) = D(i,j)$$

$$1 \leq i \leq n, 1 \leq j \leq m$$

Procedure starts by calculating all the fields with small indexes and then progressively continues to calculate the fields with higher indexes:

$$\begin{aligned} &for\ i = 1, \ldots, n \\ &\quad for\ j = 1, \ldots, m \\ &\quad\quad M(i,j) = min(M(i-1,j), M(i,j-1), M(i,j)) + dst(a_i, b_j) \end{aligned}$$

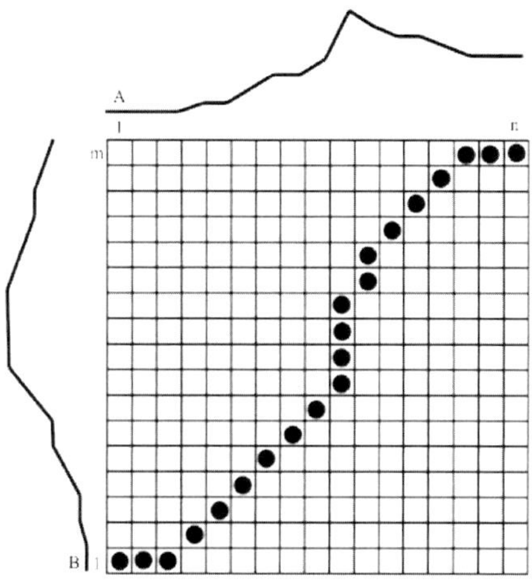

Fig. 3. Minimum distance warp path between the time series A and B

The value in the cell of a matrix M with the highest indexes $M(n,m)$ is the distance corresponding to the minimum distance warp path. A minimum distance warp path can be obtained by following cells with the smallest values from $M(n,m)$ to $M(1,1)$ (in Fig. 3 the minimum distance warp path is marked with dots).

Many attempts to speed up DTWs have been proposed [18]; these can be categorized as constraints. Constraints limit the minimum distance warp path search space by reducing the allowed warp along the time axis. The two most commonly used constraints are the Sakoe-Chiba Band [17] and Itakura Parallelogram [6], which are shown in Fig. 4.

3.4 Modification of the Algorithm for Multidimensional Classification

The DTW algorithm commonly described in the literature is suitable for aligning one-dimensional time series. This work employed a modification of the DTW, which makes it suitable for multidimensional classification.

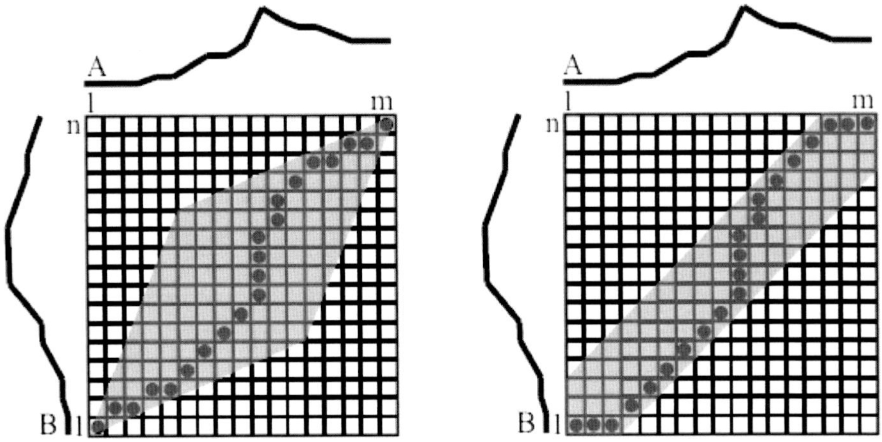

Fig. 4. Itakura Parallelogram (left) and Sakoe-Chiba Band (right) constraints. Only shaded cells are used by DTW algorithm

First, each time point of the captured time series consisting of the positions of the 12 tags coming out of the motion-capture system is transformed into angle attribute space, as defined before. The classification is then performed in the transformed space.

To align an input recording with a template recording (on which the classifier was trained), we first have to compute the matrix of local distances, $d(i, j)$, in which each element (i, j) represents the local distance between the i-th time point of the template and the input at the time j. Let C_{js} be a generic attribute-vector element relative to a template recording, and Q_{is} be the attribute-vector element relative to a new input recording to recognize, where $1 \leq s \leq N$ is the considered attribute.

For the definition of distance the Euclidean distance was used, defined as follows:

$$d_{Euc} = \sqrt{\sum_{s=1}^{N} (C_{js} - Q_{is})^2}$$

The value of the minimum global distance for the complete alignment of the DTW procedure, i.e., the final algorithm output, is found in the last column and row, $D(T_r, T_r)$. The optimal alignment can also be efficiently found by back tracing through the matrix: the alignment path starts from $D(T_r, T_r)$, then it proceeds, at each step, by selecting the cell that contains the minimum cumulative distance between those cells allowed by the alignment path constraints until $D(1, 1)$ is reached.

4 Experiments and Results

The DTW algorithm attempts to stretch and compress an input time series in order to minimize a suitably chosen distance measure from a given template. We used a nearest-neighbor classifier based on this distance measure to design the algorithm as a fall detector and a disease classifier.

The classification process considers one input time series, comparing it with the whole set of templates, computing the minimum global distance for each alignment and assuming that the input recording is in the same class of the template with which the alignment gives the smallest minimum global distance (analogous to instance-based learning).

The proposed algorithms were tested with the methodology and the data set described in the study. The 10-fold cross-validation for the 5-nearest-neighbor classifier resulted in a classification accuracy of 97.5% and 97.6% for the activities and health problems, respectively.

Table 1 and Table 2 show the confusion matrices, i.e., how many examples of a certain true class (in rows) are classified in one of five possible classes (in columns).

For the real-world cases we can use confusion matrices for three purposes:

- **False positives (false alarms):** How many can be expected using these classifiers. When in real-world use the system would report a false alarm, e.g., normal walking is classified as a health problem, the ambulance could drive to pick up the elderly person, which would result in unnecessary costs.
- **False negatives:** How many can be expected using these classifiers. False negatives could mean a potentially risky situation for the elderly person, as his/her health problem would not be recognized automatically.
- **Errors (misclassifications):** Between which health states (classes) the errors (misclassifications) occur. Consequently, we can add additional attributes to help distinguish between those particular classes. The misclassifications happened very rarely.

Table 1. Confusion matrix of k-nearest-neighbor classifier, where F=fall, UF=unconscious fall, W=walking, SS=standing/sitting, L=lying down/lying. Numbers denote the quantity of the classified examples

true class \ classified as	F	UF	W	SS	L
F	30	0	0	0	0
UF	0	30	0	0	0
W	1	0	124	0	0
SS	0	0	0	24	1
L	0	3	1	0	26

Table 2. Confusion matrix of k-nearest-neighbor classifier, where H=hemiplegia, L=pain in the leg, N=normal (healthy) walking, P=Parkinson's disease and B=Pain in the back. Numbers denote the quantity of the classified examples

true class \ classified as	H	L	N	P	B
H	42	2	1	0	0
L	0	25	0	0	0
N	1	0	24	0	0
P	0	0	0	25	0
B	0	0	0	0	21

The results show that in the proposed approach false positives/negatives are very rare, i.e., they would not cause many unnecessary ambulance costs. Since the method accurately classified most real health problems, it represents high confidence and safety for its potential use in the care of the elderly.

5 Conclusion

This study presented a generalized approach to the discovery of the chance of health problems and falls in the elderly for the purpose of prolonging the autonomous living of the elderly using time-series data mining. It is general in the sense that it does not use specific medically defined attributes but the general approach of a combined k-nearest-neighbor algorithm with multidimensional dynamic time warping. It is a two-step approach. In the first step it classifies the person's activities into five activities, including different types of falls. In the second step it classifies walking patterns into five different health states: one healthy and four unhealthy. Even though the new approach is more general and can also be used to classify other types of activities or health problems, it still achieves high classification accuracies, similar to the more specific kind of approach.

Acknowledgements. The operation leading to this study is partially financed by the European Union, European Social Fund. The authors thank Martin Tomšič, Bojan Nemec and Leon Žlajpah for their help with data acquisition and Anton Gradišek for lending his medical expertise.

References

1. Bourke, A.K., et al.: An optimum accelerometer configuration and simple algorithm for accurately detecting falls. In: Proc. BioMed 2006, pp. 156–160 (2006)
2. Bourke, A.K., O'Brien, J.V., Lyons, G.M.: Evaluation of a threshold-based tri-axial accelerometer fall detection algorithm. Gait & Posture 26, 194–199 (2007)
3. Confidence Consortium: Ubiquitous Care System to Support Independent Living, `http://www.confidence-eu.org/`
4. Craik, R., Oatis, C.: Gait Analysis: Theory and Application. Mosby-Year Book (1995)

5. eMotion: Smart motion capture system, `http://www.emotion3d.com/smart/smart.html`
6. Itakura, F.: Minimum prediction residual principle applied to speech recognition. IEEE Transactions on Acoustics, Speech and Signal Processing 23(1), 67–72 (1975)
7. Kaluža, B., Mirchevska, V., Dovgan, E., Luštrek, M., Gams, M.: An Agent-Based Approach to Care in Independent Living. In: de Ruyter, B., Wichert, R., Keyson, D.V., Markopoulos, P., Streitz, N., Divitini, M., Georgantas, N., Mana Gomez, A. (eds.) AmI 2010. LNCS, vol. 6439, pp. 177–186. Springer, Heidelberg (2010)
8. Keogh, E., Ratanamahatana, C.A.: Exact indexing of dynamic time warping. Knowl. Inf. Syst. 7(3), 358–386 (2005)
9. Lakany, H.: Extracting a diagnostic gait signature. Patt. Recognition 41, 1627–1637 (2008)
10. Luštrek, M., Kaluža, B.: Fall detection and activity recognition with machine learning. Informatica 33, 2 (2009)
11. Miskelly, F.G.: Assistive technology in elderly care. Age and Ageing 30, 455–458 (2001)
12. Moore, S.T., et al.: Long-term monitoring of gait in Parkinson's disease. Gait Posture (2006)
13. Perolle, G., Fraisse, P., Mavros, M., Etxeberria, L.: Automatic fall detection and acivity monitoring for elderly, COOP-005935 — HEBE Cooperative Research Project- CRAFT, Luxembourg (2006)
14. Pogorelc, B., Bosnić, Z., Gams, M.: Automatic recognition of gait-related health problems in the elderly using machine learning. Multimed Tools Appl. 58(2), 333-354 (2012), doi:10.1007/s11042-011-0786-1
15. Ribarič, S., Rozman, J.: Sensors for measurement of tremor type joint movements. MIDEM 37(2), 98 104 (2007)
16. Rudel, D.: Zdravje na domu na daljavo za stare osebe. Infor. Med. Slov. 13(2), 19–29 (2008)
17. Sakoe, H., Chiba, S.: Dynamic programming algorithm optimization for spoken word recognition. IEEE Transactions on Acoustics, Speech and Signal Processing 26(1), 43–49 (1978)
18. Salvador, S., Chan, P.: Toward accurate dynamic time warping in linear time and space. Intell. Data Anal. 11(5), 561–580 (2007)
19. Strle, D., Kempe, V.: MEMS-based inertial systems. MIDEM 37(4), 199–209 (2007)
20. Toyne, S.: Ageing: Europe's growing problem. BBC News, `http://news.bbc.co.uk/2/hi/business/2248531.stm`
21. Trontelj, J., et al.: Safety Margin at mammalian neuromuscular junction - an example of the significance of fine time measurements in neurobiology. MIDEM 38(3), 155–160 (2008)
22. Williams, M.E., Owens, J.E., Parker, B.E., Granata, K.P.: A new approach to assessing function in elderly people. Trans. Am. Clin. Clim. Ass. 114, 203–216 (2003)
23. Dovgan, E., Luštrek, M., Pogorelc, B., Gradišek, A., Burger, H., Gams, M.: Intelligent elderly-care prototype for fall and disease detection from sensor data. Zdrav Vestn 80, 824–831 (2011)
24. Strle, B., Mozina, M., Bratko, I.: Qualitative approximation to Dynamic Time-Warping similarity between time series data. In: Proceedings of the Workshop on Qualitative Reasoning (2009)

Temporal Logics Modeling Logical Uncertainty, Local and Global Chance Discovery

Vladimir Rybakov

School of Computing, Mathematics and Digital Technology
Manchester Metropolitan University,
Manchester M1 5GD, UK
V.Rybakov@mmu.ac.uk

Abstract. The paper develops a logical framework for studying Chance Discovery and Uncertainty. We investigate[1] a special linear temporal logic $\mathcal{TL}^{\mathcal{Z}}_{\mathcal{DU}}$ with operations UNTIL and NEXT, which combines operations of the linear temporal logic LTL, the operation for discovery (variations of chance discovery – CD) and operation for logical uncertainty. We distinguish local and global discovery using approach of temporal logic. Our main aim is to solve problems of satisfiability and decidability for $\mathcal{TL}^{\mathcal{Z}}_{\mathcal{DU}}$. Our principal result is found algorithm which checks if any given formula is true in $\mathcal{TL}^{\mathcal{Z}}_{\mathcal{DU}}$ (which implies that $\mathcal{TL}^{\mathcal{Z}}_{\mathcal{DU}}$ is decidable, and the satisfiability problem for $\mathcal{TL}^{\mathcal{Z}}_{\mathcal{DU}}$ is solvable). In the final part of the chapter we consider the case of non-linear temporal logics based on just reflexive and transitive time flow (which does not implement operations UNTIL and NEXT) with interpretations of Chance Discovery and Uncertainty. Such logics are also decidable.

Keywords: linear temporal logic, chance discovery, uncertainty, decidability algorithms, Kripke/Hintikka models.

1 Introduction

We will study modeling of Chance Discovery and Uncertainty in a logical framework based at linear temporal logic LTL and its Kripke-Hintikka models. CD is a contemporary direction in Artificial Intelligence (AI) and Computer Science (CS) (cf. Ohsawa and McBurney [10], Abe and Ohsawa [1], Abe et al. [2]) which analyzes important events with uncertain information, incomplete past data, so to say, *chance* events, where *a chance* is defined as some event which is significant for decision-making in a specified domain.

Suggested by us technique will describe one of possible ways for modeling logical properties of CD within approach based at the linear temporal logic LTL and studying interconnections of CD and logical uncertainty.

[1] This research was supported in 2011 by Engineering and Physical Sciences Research Council (EPSRC), UK, grant EP/F014406/1.

Y. Ohsawa & A. Abe (Eds.): Advances in Chance Discovery, SCI 423, pp. 177–189.
springerlink.com

To recall history, temporal logics were first suggested to specify properties of programs in the late 1970's (cf. Pnueli [11]). The most used temporal framework is the linear-time propositional temporal logic LTL, which has been extensively studied from the point of view various prospects of applications (cf. e.g. Manna and Pnueli [9], Barringer, Fisher, Gabbay and Gough [3]). Model checking for LTL formed a strong direction within logic in computer science, which uses, in particular, applications of automata theory (cf. Vardi [22]).

Temporal logics themselves can be considered as a special cases of hybrid logics, e.g. as bimodal logics with some laws imposed on the interaction of modalities to imitate the flow of time. The mathematical theory devoted to studying axiomatizations of temporal logics and developing of their semantic theory based on Kripke/Hintikka-like models and temporal Boolean algebras formed a highly technical branch in non-classical logics (cf. van Benthem [21,20], Gabbay and Hodkinson [6], Hodkinson [7]). Axiomatizations of various temporal linear logics were summarized in de Jongh et al [5].

In first part of our paper we introduce a new temporal logic $\mathcal{TL}^{\mathcal{Z}}_{\mathcal{DU}}$ combining operations of LTL, local and global operations $\mathbf{D_l}, \mathbf{D_g}$ for logical discovery, and operation **Ucn** for logical uncertainty.

Motivation for introducing this logic is its high expressive power, which combines the one from all background logics and, therefore, may model all inherent properties, and, besides, describe possible interactions of combined logical operations.

The logic is defined as the set of all formulas valid at all (Kripke/Hintikka like) models $\mathcal{Z}_C$ based on time indexed by the set $\mathcal{Z}$ of all integer numbers. We construct algorithms verifying truth and satisfiability within suggested logic, so, the problems of satisfiability and decidability for $\mathcal{TL}^{\mathcal{Z}}_{\mathcal{DU}}$ are of primary interest; we find an algorithm which recognizes theorems of $\mathcal{TL}^{\mathcal{Z}}_{\mathcal{DU}}$ (which implies that $\mathcal{TL}^{\mathcal{Z}}_{\mathcal{DU}}$ is decidable, and the satisfiability problem for $\mathcal{TL}^{\mathcal{Z}}_{\mathcal{DU}}$ is solvable).

Suggested algorithm works as follows: an arbitrary formula in the language of $\mathcal{TL}^{\mathcal{Z}}_{\mathcal{DU}}$ is, first, transformed into a rule in a special normal reduced form, which, then, is checked for validity on special models (of size efficiently bounded by the size of the rule) w.r.t special kind valuations. The general methodology of this paper is borrowed from the paper [18] (and from the research reported recently at the conference KES 2008) where logical operation for discovery was first time joined to operations of modal logics.

In the final part of the chapter we consider non-linear temporal logics based on just reflexive and transitive time flow (which do not implement operations UNTIL and NEXT) with interpretations of Chance Discovery and Uncertainty. Such logics are also decidable.

Our results may be classified as construction of mathematical models for CD and Uncertainty, development of logical theory for CD and Uncertainty; they might be useful for researchers interested in logical properties of CD and Uncertainty within AI paradigms.

2 Preliminary Discussion, Notation, Language and Semantics of $\mathcal{TL}_{\mathcal{DU}}^{\mathcal{Z}}$

Chance discovery (CD in the sequel) is a modern discipline in Information Processing (and AI in general, cf. for example Ohsawa and McBurney [10]). In general terms, CD deals with discovering chances, rare important events, which may be difficult to observe and identify. Techniques for studying CD involve determining the significance of some piece of information about an event and then, using this new knowledge in decision making (cf. [10]). These techniques primarily are based at data mining methods for finding rare but important events with knowledge management, groupware, and social psychology. In our paper we, however, will be more interested in modeling logical properties of CD, our approach will use for this purpose Kripke-Hintikka relational models.

Another component allied to CD is uncertainty (bearing in mind that the chance to discover a rare event is not certain). Usually uncertainty is understood as a lack of certainty (say, in a situation of having limited knowledge, in a case when it is impossible to exactly describe existing state or future outcome). Standard approach to formalizing uncertainty often uses measurement of uncertainty based on probability for possible states or outcomes, where probabilities are assigned to each possible state or outcome. For our approach we will use uncertainty as a new logical operation applied to statements (formulas) with the aim to find a general logical framework for studying logical properties of CD and uncertainty.

Logical system, chosen as a base for our modeling, is the Linear Temporal Logic LTL. LTL is a modal temporal logic with modalities UNTIL and NEXT referring to time. Since 1977 LTL got to be very popular instrument for modeling and studying properties of computational procedures. Using logical operations UNTIL and NEXT statements about the future of time-paths may be formalized (e.g. such as that, – a condition will eventually be true, that a condition will be true until another fact becomes true, etc). LTL was first proposed for the formal verification of computer programs by Amir Pnueli in 1977. Two main types of qualities for formal verification via LTL are safety of properties (usually a state that something bad never happens) and liveness (properties that state that something good keeps happening). We will base our modeling for CD and uncertainty on LTL semantics.

So, we start from description our sematic structures. The basic semantic objects upon which we ground our logic are the following Kripe/Hintikka models. A frame

$$\mathcal{Z}_C := \langle \bigcup_{i \in Z} C(i), R, Next, Prev \rangle$$

is a tuple, where Z is the set of all integer numbers, all $C(i)$ are disjoint nonempty sets of worlds (common term of Kripke models, which may be understood as states, events while computational runs, etc), i.e. $C(i) \cap C(j) = \emptyset$ if $i \neq j$, R is a binary linear relation for time, where

$$\forall a, b \in \bigcup_{i \in Z} C(i)(aRb) \Leftrightarrow [a \in C(i) \& b \in C(j) \& i < j] \vee \exists i \in Z[a, b \in C(i)].$$

The relation $Next$ is defined by

$$a \ Next \ b \Leftrightarrow [\exists i((a \in C(i)) \& (b \in C(i+1))].$$

Besides we will use the following definition: $a \ Prev \ b \Leftrightarrow b \ Next \ a$. The intuitive perception of these frames may be as follows. Any frame $\mathcal{Z}_C$ represents some possible unbounded (in time) computation with multi-possessors as members of $C(i)$; any $i \in Z$ (any integer number i) simulates time tick i, any $C(i)$ consists of processors (computational units) engaged in the computation in time i. Next possible interpretation of models $\mathcal{Z}_C$ might be as a representation of a network (web, local net, etc.), where $\mathcal{Z}$ models time ticks (clicks) in future for net navigation; for negative integers from Z, they represent rollback.

The language of our logic extends the language of LTL by adding new (unary) logical operations $\mathbf{D_l}$ (locally discoverable), $\mathbf{D_g}$ (globally discoverable), and **Unc** (uncertain). Also, together with the standard language of LTL (with operations **N** (next) and **U** (until)), we will use new binary logical operations $\mathbf{U}_w$ (weak until), $\mathbf{U}_s$ (strong until), **S** (since), $\mathbf{S}_w$ (weak since), $\mathbf{S}_s$ (strong since) and $\mathbf{N}^{-1}$ (previous).

The formation rules for formulas are as usual, and the intended meanings of the operations are as follows:

$\mathbf{D_l}\varphi$ means the statement φ is *locally discoverable* in the current state of the current time cluster;

$\mathbf{D_g}\varphi$ means the statement φ is *globally discoverable* in the current state;

$\mathbf{Unc}\varphi$ means the statement φ is *uncertain* in the current state of the current time cluster;

$\mathbf{N}\varphi$ has the meaning φ holds in the *next time cluster* of states (state);

$\mathbf{N}^{-1}\varphi$ means φ holds in the *previous time cluster* of states (state);

$\varphi\mathbf{U}\psi$ means: φ holds until ψ will hold;

$\varphi\mathbf{S}\psi$ φ says that since ψ was true, φ holds until now;

$\varphi\mathbf{U}_w\psi$ has the meaning φ *weakly holds* until ψ will hold;

$\varphi\mathbf{U}_s\psi$ has the meaning φ *strongly holds* until ψ will hold;

$\varphi\mathbf{S}_w\psi$ φ says that since ψ was true, φ weakly holds until now;

$\varphi\mathbf{S}_s\psi$ φ means that since ψ was true, φ strongly holds until now.

Given by arbitrary countable set of propositional letters $Prop$ and any frame $\mathcal{Z}_C$, a valuation in $\mathcal{Z}_C$ is a mapping which assigns truth values to elements of $Prop$ in $\mathcal{Z}_C$. Thus, for any $p \in Prop$, $V(p) \subseteq \mathcal{Z}_C$. We will call $\langle \mathcal{Z}_C, V \rangle$ a model (a Kripke/Hintikka model). For any such model $\mathcal{M}$, the truth values are extended from propositions of $Prop$ to arbitrary formulas as follows (for $a \in \mathcal{Z}_C$, we denote $(\mathcal{M}, a) \Vdash_V \varphi$ to say that the formula φ is true at a in $\mathcal{M}_C$ w.r.t. V). The rules are as follows:

$$\forall p \in Prop, \quad (\mathcal{M}, a) \Vdash_V p \;\Leftrightarrow\; a \in V(p); \quad (\mathcal{M}, a) \Vdash_V \neg\varphi \Leftrightarrow \; not[(\mathcal{M}, a) \Vdash_V \varphi].$$

$$(\mathcal{M}, a) \Vdash_V \varphi \wedge \psi \Leftrightarrow (\mathcal{M}, a) \Vdash_V \varphi \wedge (\mathcal{M}, a) \Vdash_V \psi;$$

For computation truth values of the logical operation *to be discoverable*, we apply the following rules:

$$(\mathcal{M}, a) \Vdash_V \mathbf{D_l}\varphi \Leftrightarrow \exists i[a \in C(i) \wedge \exists b \in C(i)(\mathcal{M}, b) \Vdash_V \varphi].$$

$$(\mathcal{M}, a) \Vdash_V \mathbf{D_g}\varphi \Leftrightarrow \exists i \exists j[a \in C(i) \wedge (i \leq j) \wedge \exists b \in C(j)(\mathcal{M}, b) \Vdash_V \varphi].$$

Thus φ is *locally discoverable* at a state of a time cluster $C(i)$ if there is a state in $C(i)$, i.e. in time i, where φ is true. In another words, CD for a statement φ may be satisfied if φ holds at least in one state of the current time cluster. For $D_g\varphi$ we look globally - in all states of all future time clusters. For computation uncertainty we use:

$$(\mathcal{M}, a) \Vdash_V \mathbf{Unc}\varphi \Leftrightarrow \exists i[a \in C(i) \wedge \exists b \in C(i)(\mathcal{M}, b) \Vdash_V \varphi \wedge$$

$$\exists c \in C(i)(\mathcal{M}, c) \Vdash_V \neg\varphi].$$

Informally, φ is *uncertain* at a state of a time cluster $C(i)$ if there are two states in $C(i)$, i.e. in time i, where φ is true at one of these states and is false at the another one. This looks as quite plausible way to express uncertainty of φ (though, clearly, one of possible ones, it could be many ways to talk about uncertainty).

Consider as an example usage of information from an internet website (say Amazon or alike). Assume someone looks for better choice of a netbook, considers Samsung N220 and checks users' reviews. As the result he/she sees that 67 percents of users classify it as very good, while 33 percents of users think it is just average. So, we may consider the statement *Samsung N220 is very good* has uncertain truth and encode this situation by **Unc**(Samsung N220 is very good).

Next, we give the rules to compute truth values of logical operations of the linear temporal logic.

$$(\mathcal{M}, a) \Vdash_V \mathbf{N}\varphi \Leftrightarrow \forall b[(a \text{ Next } b) \Rightarrow (\mathcal{M}, b) \Vdash_V \varphi];$$

$$(\mathcal{M}, a) \Vdash_V \mathbf{N^{-1}}\varphi \Leftrightarrow \forall b[(a \text{ Prev } b) \Rightarrow (\mathcal{M}, b) \Vdash_V \varphi];$$

$$(\mathcal{M}, a) \Vdash_V \varphi \mathbf{U} \psi \Leftrightarrow \exists b[(aRb) \wedge ((\mathcal{M}, b) \Vdash_V \psi) \wedge$$

$$\forall c[(aRcRb)\&\neg(bRc) \Rightarrow (\mathcal{M}, c) \Vdash_V \varphi]];$$

$$(\mathcal{M}, a) \Vdash_V \varphi \mathbf{U}_w \psi \Leftrightarrow \exists b[(aRb) \wedge ((\mathcal{M}, b) \Vdash_V \psi) \wedge$$

$$\forall c[(aRcRb)\&\neg(bRc)\&(c \in C(i)) \Rightarrow \exists d \in C(i)(\mathcal{M}, d) \Vdash_V \varphi]];$$

$$(\mathcal{M}, a) \Vdash_V \varphi \mathbf{U}_s \psi \Leftrightarrow \exists b[(aRb) \wedge b \in C(i) \wedge$$

$$\forall c \in C(i)((\mathcal{M}, c) \Vdash_V \psi) \wedge \forall c[(aRcRb)\&\neg(bRc) \Rightarrow (\mathcal{M}, c) \Vdash_V \varphi]];$$

$$(\mathcal{M}, a) \Vdash_V \varphi \mathbf{S} \psi \Leftrightarrow \exists b[(bRa) \wedge ((\mathcal{M}, b) \Vdash_V \psi) \wedge$$

$$\forall c[(bRcRa)\&\neg(cRb) \Rightarrow (\mathcal{M}, c) \Vdash_V \varphi]];$$

$$(\mathcal{M}, a) \Vdash_V \mathbf{S}_w \psi \Leftrightarrow \exists b[(bRa) \wedge ((\mathcal{M}, b) \Vdash_V \psi) \wedge$$

$$\forall c[(bRcRa)\&\neg(cRb)\&(c \in C(i)) \Rightarrow \exists d \in C(i)(\mathcal{M}, d) \Vdash_V \varphi]];$$

$$(\mathcal{M}, a) \Vdash_V \mathbf{S}_s \psi \Leftrightarrow \exists b[(aRb) \wedge b \in C(i) \wedge$$

$$\forall c \in C(i)((\mathcal{M}, c) \Vdash_V \psi) \wedge \forall c[(bRcRa)\&\neg(cRb) \Rightarrow (\mathcal{M}, c) \Vdash_V \varphi]].$$

For a Kripke structure $\mathcal{M} := \langle \mathcal{Z}_C, V \rangle$ and a formula φ, (i) φ is *satisfiable* in $\mathcal{M}$ (denotation – $\mathcal{M} \Vdash_{Sat} \varphi$) if there is a state b of $\mathcal{M}$ ($b \in \mathcal{Z}_C$) where φ is true: $(\mathcal{M}, b) \Vdash_V \varphi$. (ii) φ is *valid* in $\mathcal{M}$ (denotation – $\mathcal{M} \Vdash \varphi$) if, for any b of $\mathcal{M}$ ($b \in \mathcal{Z}_C$), the formula φ is true at b ($(\mathcal{M}, b) \Vdash_V \varphi$).

For a frame $\mathcal{Z}_C$ and a formula φ, φ is satisfiable in $\mathcal{Z}_C$ (denotation $\mathcal{Z}_C \Vdash_{Sat} \varphi$) if there is a valuation V in the frame $\mathcal{Z}_C$ such that $\langle \mathcal{Z}_C, V \rangle \Vdash_{Sat} \varphi$. φ is valid in $\mathcal{Z}_C$ (notation $\mathcal{Z}_C \Vdash \varphi$) if *not*($\mathcal{Z}_C \Vdash_{Sat} \neg\varphi$).

Definition 1. *The logic $\mathcal{TL}_{\mathcal{DU}}^{\mathcal{Z}}$ is the set of all formulas which are valid in all frames $\mathcal{Z}_C$.*

We say a formula φ is *satisfiable* iff there is a valuation V in a Kripke frame $\mathcal{Z}_C$ which makes φ satisfiable: $\langle \mathcal{Z}_C, V\rangle \Vdash_{Sat}\varphi$. Clearly, a formula φ is satisfiable iff $\neg\varphi$ is not a theorem of $\mathcal{TL}^{\mathcal{Z}}_{\mathcal{DU}}$: $\neg\varphi \notin \mathcal{TL}^{\mathcal{Z}}_{\mathcal{DU}}$, and vice versa, φ is a theorem of $\mathcal{TL}^{\mathcal{Z}}_{\mathcal{DU}}$ ($\varphi \in \mathcal{TL}^{\mathcal{Z}}_{\mathcal{DU}}$) if $\neg\varphi$ is not satisfiable. Using the operation $\mathbf{U}$ we, in the well known way, can define all standard modal and temporal operations, e.g. $\Diamond$ (possible), $\Box$ (necessary), $\mathbf{F}$ (*holds eventually*), $\mathbf{G}$ (*holds henceforth*), etc.

Our logic $\mathcal{TL}^{\mathcal{Z}}_{\mathcal{DU}}$, introduced above, is more expressive compared to standard LTL. We handle variations of the operation Until and Since: the new temporal operations $\mathbf{U}_s$ and $\mathbf{U}_w$ bring new unique features to the language. For instance the formula $\Box_w\varphi := \neg(\top\mathbf{U}_s\neg\varphi)$ codes *weak necessity*, it says that in any future time cluster $C(i)$ there is a state where φ is true. So, this formula way code the *non-vacuity* for a task p computation: in any tick in future at least one possessor unit works with p. The formula $(\neg\varphi\mathbf{U}_w\Box\varphi)\wedge\Diamond\Box\varphi$ codes that there is a minimal time point i since which φ holds in all states of all future time clusters, but before the time point i the formula φ is false in a state of any time cluster. Such properties are problematic to be expressed in standard modal or temporal operations.

Used by us operations to handle CD and Uncertainty also show general interdependencies in chosen approach. For instance,

$$\mathbf{Unc}\varphi \ \rightarrow \ \mathbf{D_l}\varphi \ \mathcal{TL}^{\mathcal{Z}}_{\mathcal{DU}},$$

$$\mathbf{D_l}\varphi \wedge \mathbf{D_l}\neg\varphi \ \rightarrow \ \mathbf{Unc}\varphi \ \mathcal{TL}^{\mathcal{Z}}_{\mathcal{DU}}.$$

$$\mathbf{D_g}\varphi \Leftrightarrow T\mathbf{U}\varphi \in \mathcal{TL}^{\mathcal{Z}}_{\mathcal{DU}}.$$

3 Decidability Algorithm for $\mathcal{TL}^{\mathcal{Z}}_{\mathcal{DU}}$

For any logical system, one of most fundamental questions is the decidability problem: if there is an algorithm computing theorems of this logic. We address this problem to our logic $\mathcal{TL}^{\mathcal{Z}}_{\mathcal{DU}}$. The basic technique we use is based on the reduction of formulas in the language of $\mathcal{TL}^{\mathcal{Z}}_{\mathcal{DU}}$ to special inference rules and the verification of the validity these rules in frames $\mathcal{Z}_C$. This approach uses techniques to handle inference rules from [12] – [19] (where [19] solves decidability of LTL w.r.t. admissibility and again decidability of LTL itself).

Recall, a (sequential) (inference) rule is a relation

$$\mathbf{r} := \frac{\varphi_1(x_1,\ldots,x_n),\ldots,\varphi_l(x_1,\ldots,x_n)}{\psi(x_1,\ldots,x_n)},$$

where

$$\varphi_1(x_1,\ldots,x_n),\ldots,\varphi_l(x_1,\ldots,x_n) \text{ and } \psi(x_1,\ldots,x_n)$$

are formulas constructed out of letters $x_1,\ldots,x_n$. The letters $x_1,\ldots,x_n$ are the variables of $\mathbf{r}$, we use the notation $x_i \in Var(\mathbf{r})$.

Definition 2. *A rule* **r** *is said to be* **valid** *in a Kripke model* $\langle \mathcal{Z}_C, V\rangle$ *(notation* $\mathcal{Z}_C \Vdash_V \mathbf{r}$*) if*

$$[\forall a\ ((\mathcal{Z}_C, a) \Vdash_V \bigwedge_{1\leq i\leq l} \varphi_i)] \Rightarrow \forall a\ ((\mathcal{Z}_C, a) \Vdash_V \psi).$$

Otherwise we say **r** *is* **refuted** *in* $\mathcal{Z}_C$*, or* **refuted in** $\mathcal{Z}_C$ **by** V*, and write* $\mathcal{Z}_C \nVdash_V \mathbf{r}$*. A rule* **r** *is* **valid** *in a frame* $\mathcal{Z}_C$ *(notation* $\mathcal{Z}_C \Vdash \mathbf{r}$*) if, for any valuation* V*,* $\mathcal{Z}_C \Vdash_V \mathbf{r}$*.*

For any formula φ we can convert it into the rule $x \rightarrow x/\varphi$ and employ a technique of reduced normal forms for inference rules as follows. Evidently,

Lemma 1. *A formula* φ *is a theorem of* $\mathcal{TL}_{\mathcal{DU}}^{\mathcal{Z}}$ *iff the rule* $(x \rightarrow x/\varphi)$ *is valid in any frame* $\mathcal{Z}_C$*.*

A rule **r** is said to be in *reduced normal form* if $\mathbf{r} = \varepsilon/x_1$ where

$$\varepsilon := \bigvee_{1\leq j\leq l} (\bigwedge_{1\leq i,k\leq n, i\neq k} [x_i^{t(j,i,0)} \wedge (\mathbf{D_l} x_i)^{t(j,i,1)} \wedge (\mathbf{D_g} x_i)^{t(j,i,2)} \wedge$$

$$(\mathbf{Unc} x_i)^{t(j,i,3)} \wedge (\mathbf{N} x_i)^{t(j,i,4)} \wedge (\mathbf{N^{-1}} x_i)^{t(j,i,5)} \wedge (x_i \mathbf{U} x_k)^{t(j,i,k,0)} \wedge$$

$$(x_i \mathbf{U}_w x_k)^{t(j,i,k,1)} \wedge\ (x_i \mathbf{U}_s x_k)^{t(j,i,k,2)} \wedge (x_i \mathbf{S} x_k)^{t(j,i,k,3)} \wedge$$

$$(x_i \mathbf{S}_w x_k)^{t(j,i,k,4)} \wedge (x_i \mathbf{S}_s x_k)^{t(j,i,k,5)}]),$$

where all x_s are certain letters (variables), $t(j,i,z), t(j,i,k,z) \in \{0,1\}$ and, for any formula α above, $\alpha^0 := \alpha$, $\alpha^1 := \neg\alpha$.

Definition 3. *Given a rule* $\mathbf{r_{nf}}$ *in reduced normal form,* $\mathbf{r_{nf}}$ *is said to be a normal reduced form for a rule* **r** *iff, for any frame* $\mathcal{Z}_C$*,* $\mathcal{Z}_C \Vdash \mathbf{r} \Leftrightarrow \mathcal{Z}_C \Vdash \mathbf{r_{nf}}$ *.*

By verbatim following to proof of Lemma 3.1.3 and Theorem 3.1.11 in [13] we obtain

Theorem 1. *There exists an algorithm running in (single) exponential time, which, for any given rule* **r***, constructs its normal reduced form* $\mathbf{r_{nf}}$*.*

Decidability of $\mathcal{TL}_{\mathcal{DU}}^{\mathcal{Z}}$ will follow (by Lemma 1) if we find an algorithm recognizing rules in reduced normal form which are valid in all frames $\mathcal{Z}_C$. We need one more construction on Kripke frames given below. For any frame $\mathcal{Z}_C$ and some integer numbers k_1, m_1, k_2, m_2, where $m_2 > k_2 > k_1 + 3, k_1 > m_1$ we construct the frame $\mathcal{Z}_C(k_1, m_1, k_2, m_2)$ from $\mathcal{Z}_C$ as follows. $\mathcal{Z}_C(k_1, m_1, k_2, m_2) := \langle \bigcup_{m_1 \leq i \leq m_2} C(i), R, Next\rangle$, where R is the accessibility relation from $\mathcal{Z}_C$ extended by pairs (x, y), where $x \in C(i), y \in C(j)$ and $i, j \in [m_1, k_1]$, or $i, j \in [k_2, m_2]$.

Relations *Next* and *Prev* are taken from $\mathcal{Z}_C$ and extended by $\forall a \in C(m_2)\forall b \in C(k_2)(a\ Next\ b = true)$; $\forall a \in C(m_2)\forall b \in C(k_2)(b\ Prev\ a = true)$; $\forall a \in C(m_1)\forall b \in C(k_1)(a\ Prev\ b = true)$; $\forall a \in C(m_1)\forall b \in C(k_1)(b\ Next\ a = true)$. For any valuation V of letters from a formula φ in $\mathcal{Z}_C(k_1, m_1, k_2, m_2)$ the truth value of φ can be defined at elements of $\mathcal{Z}_C(k_1, m_1, k_2, m_2)$ by the rules similar to the ones given for the frames $\mathcal{Z}_C$ above (just in accordance with the meaning of logical operations). Due to limitations on the length of the paper we omit detail description of these rules.

Lemma 2. *A rule* $\mathbf{r_{nf}}$ *in reduced normal form is refuted in a frame* $\mathcal{Z}_C$ *iff* $\mathbf{r_{nf}}$ *can be refuted in a frame* $\mathcal{Z}_C(k_1, m_1, k_2, m_2)$ *by a valuation* V *of* special kind, *where the size of the frame* $\mathcal{Z}_C(k_1, m_1, k_2, m_2)$ *is triple exponential in* $\mathbf{r_{nf}}$.

From Theorem 1, Lemma 1 and Lemma 2 we derive

Theorem 2. *The logic* $\mathcal{TL}_{\mathcal{DU}}^{\mathcal{Z}}$ *is decidable. The algorithm for checking a formula to be a theorem of* $\mathcal{TL}_{\mathcal{DU}}^{\mathcal{Z}}$ *consists in verification of validity rules in reduced normal form at frames* $\mathcal{Z}_C(k_1, m_1, k_2, m_2)$ *of size triple-exponential in the size of reduced normal forms w.r.t. valuations of special kind.*

It is possible also to apply the technique from this paper to weakened versions of the logic $\mathcal{TL}_{\mathcal{DU}}^{\mathcal{Z}}$, say with omitted strong or weak versions of the operations **U** or **S**, with omitted **N** or $\mathbf{N^{-1}}$ and to obtain similar results about decidability.

4 Case of Non-linear Temporal Logics

In this section we consider a framework for CD based on non-linear temporal logic. In this case logics have no analogues for operations UNTIL and NEXT; therefore it is much easier to get deciding algorithms.

We start from a recall of basic definitions. The language of standard temporal logics consists of the language of Boolean logic extended by two unary temporal operations $\Diamond^+$ (will be in the future) and $\Diamond^-$ (was in the past). $\Diamond^+\varphi$ to be read *there is a future state where* φ *is true*, $\Diamond^-\varphi$ means *there was a state in past where* φ *was true*. Here $\Diamond^+$ and $\Diamond^-$ to be understood as a chance to discover in future and a chance to discover in past.

We extend the language by unary temporal operation $\Diamond_l$ for local possibility: $\Diamond_l\varphi$ has meaning: φ is possible locally. The operation $\Diamond_l$ is needed to define our approach to model *uncertainty* in temporal logic.

A Kripke/Hintikka frame is a pair $\mathcal{F} := \langle F, R\rangle$, where F is the base of $\mathcal{F}$ – a non-empty set, and R is a binary (accessibility by time) relation on F. $|\mathcal{F}| := F$, $a \in \mathcal{F}$ is a denotation for $a \in |\mathcal{F}|$. In this paper we consider only reflexive and transitive temporal logics, so R in the sequel is always reflexive and transitive. In what follows, R^{-1} is the relation converse to R. If, for a set of propositional letters P, a valuation V of P in $|\mathcal{F}|$ is defined, i.e. $V : P \to 2^F$, in other words, $\forall p \in P(V(p) \subseteq F)$, then the tuple $\mathcal{M} := \langle \mathcal{F}, V\rangle$ is called a Kripke/Hintikka model (structure). The truth values of formulas are defined at elements of $\mathcal{F}$ by the following rules:

$$\forall p \in Prop,\ \forall a \in \mathcal{F},\quad (\mathcal{F}, a) \Vdash_V p \Leftrightarrow a \in V(p);$$

$$(\mathcal{F}, a) \Vdash_V \varphi \wedge \psi \Leftrightarrow (\mathcal{F}, a) \Vdash_V \varphi \text{ and } (\mathcal{F}, a) \Vdash_V \psi;$$

$$(\mathcal{F}, a) \Vdash_V \varphi \vee \psi \Leftrightarrow (\mathcal{F}, a) \Vdash_V \varphi \text{ or } (\mathcal{F}, a) \Vdash_V \psi;$$

$$(\mathcal{F}, a) \Vdash \varphi \to \psi \Leftrightarrow \neg[(\mathcal{F}, a) \Vdash_V \varphi] \text{ or } (\mathcal{F}, a) \Vdash_V \psi;$$

$$(\mathcal{F}, a) \Vdash_V \neg\varphi \Leftrightarrow \neg[(\mathcal{F}, a) \Vdash_V \varphi];$$

$$(\mathcal{F}, a) \Vdash_V \Diamond^+ \varphi \Leftrightarrow \exists b \in \mathcal{F}((aRb) \wedge (\mathcal{F}, b) \Vdash_V \varphi);$$

$$(\mathcal{F}, a) \Vdash_V \Diamond^- \varphi \Leftrightarrow \exists b \in \mathcal{F}((bRa) \wedge (\mathcal{F}, b) \Vdash_V \varphi).$$

$$(\mathcal{F}, a) \Vdash_V \Diamond_l \varphi \Leftrightarrow \exists b \in \mathcal{F}((aRb) \wedge (bRa) \wedge (\mathcal{F}, b) \Vdash_V \varphi).$$

For any $a \in \mathcal{F}$, $Val_V(a) := \{p_i \mid p_i \in P, (\mathcal{F}, a) \Vdash_V p_i\}$. For any formula φ, $V(\varphi) := \{a \mid a \in \mathcal{F}, (\mathcal{F}, a) \Vdash_V \varphi\}$. Some more abbreviations will be used in the sequel: for any formula φ, $\Box^+ \varphi := \neg\Diamond^+\neg\varphi$, $\Box^- \varphi := \neg\Diamond^-\neg\varphi$. For $a \in \mathcal{F}$, $C(a) := \{b \mid (aRb) \wedge (bRa)\}$, i.e. $C(a)$ is the cluster containing a.

Definition 4. *For a Kripke-Hintikka structure $\mathcal{M} := \langle \mathcal{F}, V \rangle$ and a formula φ, φ is* true *in $\mathcal{M}$ (denotation – $\mathcal{M} \Vdash \varphi$) if $\forall a \in \mathcal{F}$ $(\mathcal{F}, a) \Vdash_V \varphi$. $\mathcal{F} \Vdash_V \varphi \Leftrightarrow \forall w \in \mathcal{F}((\mathcal{F}, w) \Vdash_V \varphi)$.*

Our suggestion to interpret uncertainty is as follows: we introduce the following new operation U definable in offered language:

$$U\varphi := \Diamond_l^+ \varphi \wedge \Diamond_l^+ \neg\varphi.$$

The meaning of $U\varphi$ is: *the statement φ is uncertain.* It seems this approach models very well the intuitive understanding of uncertainty: truth $U\varphi$ means that *today* – in current time cluster – both φ and $\neg\varphi$ may be true, what concludes that φ is *uncertain* (at least from the current temporal viewpoint).

Our interpretation for *Chance Discovery (CD)* is based on possibility to discover an event in future (local future) or past. So, $\Diamond^+ \varphi$ says that the truth of φ is discoverable (in future computation, via web linking etc.), $\Diamond_l^+ \varphi$ says about possibility to discover that φ is true in local future, and $\Diamond^- \varphi$ says about past.

Definition 5. *For a class $\mathcal{K}$ of frames, the logic $\mathcal{L}(\mathcal{K})$ generated by $\mathcal{K}$ is the set of all formulas which are true in all models based on frames from $\mathcal{K}$.*

The temporal logic $T_{S4,U}$ is the logic $\mathcal{L}(\mathcal{K}_{r+tr})$, where $\mathcal{K}_{r+tr}$ is the class of all reflexive and transitive frames, in described above language; T_{S4} is the logic $\mathcal{L}(\mathcal{K}_{r+tr})$ in the language with omitted $\Diamond_l$.

For any logic $\mathcal{L}(\mathcal{K})$, a formula φ is a *theorem* of $\mathcal{L}(\mathcal{K})$ if $\varphi \in \mathcal{L}(\mathcal{K})$, φ is satisfiable in $\mathcal{K}$ if, for some valuation V in some frame $\mathcal{F} \in \mathcal{K}$, φ is true w.r.t. V at some world of $\mathcal{F}$.

Definition 6. *A logic $\mathcal{L}$ has the finite model property (fmp in the sequel) if $\mathcal{L} = \mathcal{L}(\mathcal{K})$, where $\mathcal{K}$ is a class of finite frames.*

It is well known that $T_{S4} = \mathcal{L}(\mathcal{K}_{r+tr,f})$, where $\mathcal{K}_{r+tr,f}$ is the class of all finite frames from $\mathcal{K}_{r+tr}$, so, T_{S4} has fmp. By a bit refined technique similar to standard filtration tools it is simple to show that $T_{S4,U}$ also has fmp. Besides, it may be shown that we can achieve fmp with effective bound for the size of the satisfying finite models. Therefore we obtain

Theorem 3. *Temporal logic $T_{S4,U}$ is decidable. There is an algorithm for verification if a formulas is a theorem of $T_{S4,U}$ (and consequently for satisfiability verification of formulas in $T_{S4,U}$).*

5 Conclusion, Future Work

Our research develops a technique (based at non-classical mathematical logic) for studying CD and Uncertainty. Our main aim is to find an efficient framework correctly describing logical properties of these notions. In particular, we find an algorithm which can compute true logical laws of the proposed logic $\mathcal{TL}_{\mathcal{DU}}^{\mathcal{Z}}$ based on the linear temporal logic LTL. This algorithm is based at model theoretic technique applied to special Kripke/Huntikka models and usage of technique to transform formulas to special sequents (inference rules) and further transformation them to special reduced forms. The suggested algorithm (as usual in similar cases) can verify satisfiability of formulas.

There are many prospective avenues for future investigation of logic $\mathcal{TL}_{\mathcal{DU}}^{\mathcal{Z}}$ and its variants. For instance, the question of finding axiomatization for $\mathcal{TL}_{\mathcal{DU}}^{\mathcal{Z}}$ is open yet. The problem of computation admissibility of inference rules in $\mathcal{TL}_{\mathcal{DU}}^{\mathcal{Z}}$ is also not investigated. Another interesting problem concerns complexity issues and possible ways of refining the complexity bounds in the suggested algorithm. The problem of describing bases for rules admissible in $\mathcal{TL}_{\mathcal{DU}}^{\mathcal{Z}}$ is also open.

We showed (in Theorem 3) that the temporal logic $T_{S4,U}$ itself is decidable, but solutions of admissibility problem for neither $T_{S4,U}$ nor T_{S4} itself are known up to date.

Our technique and obtained results might be useful for scientists interested in logical properties of CD and Uncertainty and their applications in AI. Tools developed in our paper seem to be promising for investigation properties of logical operations in other similar logics originating in AI, and we presume that effective algorithms verifying truth and satisfiability may be constructed.

References

1. Abe, A., Ohsawa, Y. (eds.): Readings in Chance Discovery. International Series on Advanced Intelligence (2005)
2. Abe, A., Hagita, N., Furutani, M., Furutani, Y., Matsuoka, R.: Exceptions as Chance for Computational Chance Discovery. KES Journal (2), 750–757 (2008)
3. Barringer, H., Fisher, M., Gabbay, D., Gough, G.: Advances in Temporal Logic. Applied logic series, vol. 16. Kluwer Academic Publishers, Dordrecht (1999)
4. Crestani, F., Lalmas, M.: Logic and Uncertainty in Information Retrieval. In: Agosti, M., Crestani, F., Pasi, G. (eds.) ESSIR 2000. LNCS, vol. 1980, p. 179. Springer, Heidelberg (2001), ISBN:3-540-41933-0
5. de Jongh, D., Veltman, F., Verbrugge, R.: Completeness by construction for tense logics of linear time. In: Troelstra, A.S., Visser, A., van Benthem, J., Veltman, F. (eds.) Liber Amicorum for Dick de Jongh. Institute of Logic, Language and Computation, Amsterdam (2004), `http://www.illc.uva.nl/D65/`
6. Gabbay, D.M., Hodkinson, I.M.: An axiomatisation of the temporal logic with Until and Since over the real numbers. Journal of Logic and Computation 1, 229–260 (1990)
7. Hodkinson, I.: Temporal Logic: Mathematical Foundations and Computational Aspects. In: Gabbay, D.M., Reynolds, M.A., Finger, M. (eds.) Temporal Logic and Automata. ch. II, vol. 2, pp. 30–72. Clarendon Press, Oxford (2000)
8. Elvang-Goransson, M., Krause, P.J., Fox, J.: Acceptability of Arguments as Logical Uncertainty. In: Moral, S., Kruse, R., Clarke, E. (eds.) ECSQARU 1993. LNCS, vol. 747, pp. 85–90. Springer, Heidelberg (1993)
9. Manna, Z., Pnueli, A.: Temporal Verification of Reactive Systems: Safety. Springer (1995)
10. Ohsawa, Y., McBurney, P. (eds.): Chance Discovery. Advanced Information Processing. Springer (2003)
11. Pnueli, A.: The Temporal Logic of Programs. In: Proc. of the 18th Annual Symp. on Foundations of Computer Science, pp. 46–57. IEEE (1977)
12. Rybakov, V.V.: Rules of Inference with Parameters for Intuitionistic logic. Journal of Symbolic Logic 57(3), 912–923 (1992)
13. Rybakov, V.V.: Admissible Logical Inference Rules. Studies in Logic and the Foundations of Mathematics, vol. 136. Elsevier Sci. Publ., North-Holland (1997)
14. Rybakov, V.V.: Construction of an Explicit Basis for Rules Admissible in Modal System S4. - Mathematical Logic Quarterly 47(4), 441–451 (2001)
15. Rybakov, V.V.: Logical Consecutions in Discrete Linear Temporal Logic. Journal of Symbolic Logic 70(4), 1137–1149 (2005)
16. Rybakov, V.V.: Logical Consecutions in Intransitive Temporal Linear Logic of Finite Intervals. Journal of Logic Computation 15(5), 633–657 (2005)
17. Rybakov, V.: Until-Since Temporal Logic Based on Parallel Time with Common Past. Deciding Algorithms. In: Artemov, S., Nerode, A. (eds.) LFCS 2007. LNCS, vol. 4514, pp. 486–497. Springer, Heidelberg (2007)
18. Rybakov, V.: Logic of Discovery in Uncertain Situations– Deciding Algorithms. In: Apolloni, B., Howlett, R.J., Jain, L. (eds.) KES 2007, Part II. LNCS (LNAI), vol. 4693, pp. 950–958. Springer, Heidelberg (2007)
19. Rybakov, V.: Linear Temporal Logic with Until and Next, Logical Consecutions - Annals of Pure and Applied Logic 155(1), 32–45 (2008)

20. van Benthem, J., Bergstra, J.A.: Logic of Transition Systems. Journal of Logic, Language and Information 3(4), 247–283 (1994)
21. van Benthem, J.: The Logic of Time. Kluwer (1991)
22. Vardi, M.: Reasoning about the past with two-way automata. In: Larsen, K.G., Skyum, S., Winskel, G. (eds.) ICALP 1998. LNCS, vol. 1443, pp. 628–641. Springer, Heidelberg (1998)

Discovering Probabilistic Models of Pilot Behavior from Aircraft Telemetry Data

Kirk Ogaard and Ronald Marsh

Department of Computer Science, University of North Dakota, Grand Forks, North Dakota, USA
{ogaard,rmarsh}@cs.und.edu

Abstract. A mobile ground-based sense-and-avoid system for Unmanned Aircraft System (UAS) operations was developed by the University of North Dakota. This system detected proximate aircraft with various sensor systems, including a 2D radar and an Automatic Dependent Surveillance – Broadcast (ADS-B) receiver. Information about those aircraft was then displayed to UAS operators with customized visualization software. Its risk mitigation subsystem was designed to estimate the current risk of midair collision for UAS operations below 18,000 feet MSL. However, accurate probabilistic models for the behavior of pilots of manned aircraft flying at these altitudes were needed before this subsystem could be implemented. In this paper the authors present the results of data mining a Flight Data Monitoring (FDM) data set from a consecutive 9 month period in 2011. Arbitrarily complex subpaths were discovered from the data set using an ant colony algorithm. Then, probabilistic models were data mined from those subpaths using extensions of the Genetic K-Means and Expectation-Maximization algorithms.

Keywords: Unmanned aircraft, collision avoidance, chance discovery, ant colony algorithms.

1 Introduction

Unmanned Aircraft (UAs) have many unique advantages over manned aircraft. With no pilot on board, UAs can be used in dangerous situations or used for repetitive, monotonous operations. With no pilot to account for in their design, UAs can be designed to carry more fuel, thus extending their maximum flight times. With its smaller size and weight, a UA also benefits from lower manufacturing and operating costs.

However, flying UAs in the U.S. National Airspace System (NAS) can be problematic as it is unclear whether or not Federal Aviation Regulations even apply to unmanned aircraft [1]. There are also safety guidelines that Unmanned Aircraft System (UAS) operators must follow. A previous analysis of ground and midair collision risks found that most UAS operations would not meet the target levels of safety established by the Federal Aviation Administration (FAA) without the incorporation of a risk mitigation strategy.

Y. Ohsawa & A. Abe (Eds.): Advances in Chance Discovery, SCI 423, pp. 191–214.
springerlink.com

Given the regulatory difficulties of flying UAs in the NAS, the University of North Dakota (UND) is identifying airspace within the state where organizations interested in developing UASs can test and operate their systems without needing onboard sense-and-avoid capabilities. UND hopes to provide an interim risk mitigation strategy to allow UAS research and development outside restricted airspace. This will facilitate the development of appropriate regulations for UAS operations in the NAS.

Current regulatory problems in the U.S. exist regarding UAs flying in the NAS. For example, many federal agencies could benefit from UAS operations in the NAS, including the U.S. Coast Guard, the U.S. Customs and Border Protection, and the U.S. Department of Agriculture. However, the FAA will not permit UASs to operate in the same airspace as manned aircraft until those UAS operations are demonstrated to have an equivalent level of safety to manned aircraft operating in the NAS.

Collision risk is defined as the potential for a UA to damage people and/or property as a result of a midair collision or ground collision. Furthermore, the risk of a particular collision is determined by the probability of the collision occurring and the amount of damage it would likely cause to people and/or property. The type of probabilistic models presented in this paper could only be used to calculate the risk of a midair collision between a UA and a manned aircraft. Any probabilistic models for calculating the risk of a UA colliding with the ground would be based on entirely different factors.

Current FAA regulations only permit UAS operations in the NAS on a case-by-case basis via Certificates of Authorization (COAs). A COA designates a restricted airspace for an authorized organization to fly a specific type of UA. While the authorized organization is flying its UA, manned aircraft are not permitted to enter the restricted airspace designated in the COA. According to FAA guidelines [2], severe collision events (i.e. collisions resulting in numerous casualties and/or annihilation of aircraft) for manned aircraft should occur less frequently than 10^{-9} occurrences per flight hour. Since current FAA regulations require the safety levels for UAS operations in the NAS to be equivalent to the safety levels for manned aircraft systems, UAS operations in the NAS would need to result in less than 10^{-9} severe collision events per flight hour.

Since many different types of UAs exist, each with different performance characteristics, categorizing UAs based on the altitude ranges which they operate at most frequently—low altitudes, medium altitudes, or high altitudes—is suggested [1]. Specific collision-avoidance strategies are recommended for UASs operating in each of the 3 altitude ranges. Also, two general strategies are identified for avoiding midair collisions between a UA and a manned aircraft: 1) maintain adequate horizontal and vertical separation between the UA and proximate aircraft or 2) if adequate separation cannot be maintained, the UA must activate an airborne sense-and-avoid system.

A ground-based sense-and-avoid system, such as GPAR-RMS [3], which consistently maintained adequate separation between the UA and proximate aircraft, made an airborne sense-and-avoid system unnecessary. Probabilistic models for manned aircraft, such as those discovered with the algorithms discussed in this paper, could have enabled GPAR-RMS to estimate the risk of a midair collision between the UA and other manned aircraft in the surrounding airspace. When provided with a current and accurate estimate of the risk of a midair collision, the Range Safety Operator (RSO) could have then immediately landed the UA if the estimated risk of a midair collision was too high.

Although many commercial aircraft are equipped with collision avoidance technology based on Automatic Dependant Surveillance – Broadcast (ADS-B) and/or Traffic Collision Avoidance System (TCAS), neither of these technologies are currently available for most UAs [1]. Furthermore, most General Aviation (GA) aircraft in the U.S. do not currently have ADS-B or TCAS capability. Thus, other systems for avoiding midair collisions between a UA and proximate aircraft (e.g. ground-based sense-and-avoid systems) need to be developed.

GPAR-RMS was designed to be an extension of the ground-based observer. GPAR-RMS integrated aircraft telemetry data (e.g. latitude, longitude, and altitude) from ground-based sensors such as ADS-B, 2D radar, and UAS Ground Control Stations (GCSs) and displayed those data on two types of Information Display Systems (IDSs). Aircraft telemetry data obtained from the various sensors were acquired and fused by the Sensor Data Fusion (SDF) subsystem. The SDF subsystem then multicast those data over a local area network to the Range Control Center IDS (RCC IDS) and Ground Observer IDS (GO IDS). Since the SDF subsystem had to acquire and present the applicable sensor data in real time, a multithreaded software system was designed which ran on high speed, multi-core servers. The system architecture for GPAR-RMS is shown in Fig.1.

The IDSs included a high-resolution wide-screen RCC IDS and one, or more, high-resolution wide-screen GO IDSs. The RCC IDS, which was modeled after existing Air Traffic Control (ATC) and Traffic Information Service – Broadcast (TIS-B) display systems, displayed the georeferenced Global Positioning System (GPS) positions of all proximate aircraft, the georeferenced boundaries of any COAs, the georeferenced positions of ground-based hazards, and current weather information from a weather station. The intent of the RCC IDS was to provide a RSO with a top-down view of the airspace, providing another level of safety and redundancy.

The GO IDS, which was modeled after existing Flight Information Service – Broadcast (FIS-B) moving map display systems, displayed the georeferenced positions of all proximate aircraft, the georeferenced boundaries of any COAs, and current weather information from a weather station. The GO IDS provided a top-down orthogonal view of the airspace around the UA, providing an unambiguous check of the UA's position compared to that of proximate aircraft. Should the current GPS position of the UA have become unavailable to the UA pilot, the GO IDS could have been used to help guide the UA back to a predetermined recovery point.

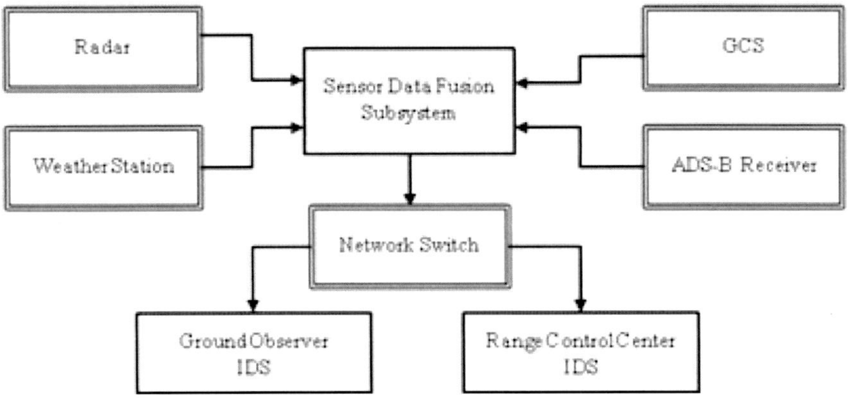

Fig. 1. The system architecture for GPAR-RMS

An unfinished task was the development of the Risk Mitigation (RM) subsystem. The RM subsystem was intended for the chance discovery of the possibility of a midair collision between the UA and proximate aircraft. The problem of estimating the total risk of a midair collision for a UA flying in a specific airspace is algorithmically complex. In order to estimate this value, the algorithm first needs estimates of the risk of every possible midair collision that could occur between the UA and any proximate aircraft. Furthermore, the RM subsystem would have needed to estimate the total risk of a midair collision for the UA in near real time. To estimate the risk of a midair collision between the UA and another aircraft, the algorithm would have needed the probability for every maneuver the pilot could perform with the aircraft during the next minute. Obviously, the pilot controlling the UA from the GCS would have known the maneuvers the UA would perform during the next minute.

Although the basic maneuvers (e.g. straight ascents/descents, ascending/descending turns, and level turns) that a pilot can perform in an aircraft are known, the flight path a pilot chooses for an aircraft can be composed of any combination of these basic maneuvers. Also, many variations for each basic maneuver exist. The pilot of an aircraft may, for instance, perform a level turn at different rates, such as 2 per second or 3 per second. Furthermore, according to FAA regulations [4], pilots flying aircraft under Visual Flight Rules (VFRs) in Class E airspace (i.e. at altitudes below 18,000 feet MSL) are not required to file flight plans. Hence the probability of the pilot of an aircraft performing any possible maneuver in Class E airspace needs to be discovered.

Our previous approach to this problem [5] was to classify pilot maneuvers extracted from ADS-B data sets according to a static classification scheme of basic maneuvers typically performed by pilots. Once classified, these subpaths were then data mined using an extension of the Genetic K-Means Algorithm (GKA) [7,10]. Data mining aircraft telemetry data sets, such as ADS-B or Flight Data Monitoring (FDM) data sets, using hybrid genetic algorithms such as GKA, can aid in discovering probabilistic models of pilot behavior in manned aircraft. An FDM data set contains very accurate data concerning the flight paths of FDM-equipped aircraft over a specific time period. Accurate estimations of the probabilities of frequently occurring maneuvers could be mined from a large FDM data set. These estimated probabilities could then be used to fully implement a ground-based sense-and-avoid system for UASs, such as GPAR-RMS.

2 Background

Unsupervised learning algorithms are appropriate for data mining problems where the classes for the data are not known *a priori*, i.e. there is no labeled training data set. In unsupervised learning algorithms, the task is to classify the data into a predetermined number of classes based on some measure of similarity between data points [10]. GKA is a hybrid algorithm which combines a genetic algorithm with the K-Means Algorithm (KMA). Since GKA uses a special class of genetic algorithms, it will eventually converge to the globally optimal solution [8],[9]. Although GKA does not directly address the cluster initialization problem of KMA, its use of a population of many candidate solutions (instead of the single candidate solution in KMA) may reduce its sensitivity to initial conditions.

The Incremental Genetic K-Means Algorithm (IGKA) and Hybrid Genetic K-Means Algorithm (HGKA) are described in [10]. IGKA and HGKA try to improve the run-time efficiency of GKA with various techniques. For example, when invalid candidate solutions are generated, they are assigned the lowest possible fitness values. Thus, invalid candidate solutions will not be selected for reproduction, and will be eliminated from succeeding generations. In contrast, GKA explicitly scans for invalid candidate solutions, which is more CPU-intensive.

The Genetic Algorithm Initialized K-Means (GAIK) algorithm is a hybrid algorithm which addresses the cluster initialization problem of KMA [11]. The GAIK hybrid algorithm uses a genetic algorithm to find initial cluster centroids which are close to global extrema. Then, KMA is run with these optimized initial centroids. This reduces the extreme sensitivity of KMA to initial conditions. However, the hybridization of KMA with a genetic algorithm makes it more CPU-intensive. Also, the GAIK hybrid algorithm only uses a genetic algorithm for finding optimal initial cluster centroids. It does not combine a genetic algorithm with KMA like GKA does. Thus, the GAIK hybrid algorithm is susceptible to converging to solutions that are only locally optimal.

Like the GAIK algorithm, the Partition-Based Genetic Algorithm Initialized K-Means (PGAIK) algorithm is a hybrid algorithm which addresses the cluster initialization problem of KMA [12]. The PGAIK hybrid algorithm uses a genetic algorithm to find optimal initial cluster centroids, and then runs KMA using those optimized initial centroids. The PGAIK hybrid algorithm partitions the data set into K subsets. It then selects one initial cluster centroid from each subset. This avoids the case where all the initial cluster centroids are very close to each other, which produces a suboptimal distribution of data points among the clusters. The PGAIK hybrid algorithm was also shown to produce more compact clusters than the GAIK algorithm. However, like the GAIK algorithm, the PGAIK hybrid algorithm is more CPU-intensive than KMA. Furthermore, the PGAIK hybrid algorithm does not retain the most optimal candidate solution it discovers during its execution. Thus, it is susceptible to converging to solutions that are only locally optimal.

An extension of the Expectation-Maximization (EM) algorithm [14], [15], [16], [17] for parallel execution on NVIDIA's Compute Unified Device Architecture (CUDA) is described in [13]. The run-time performance of their parallelized EM algorithm improved when the number of available Graphics Processing Units (GPUs) increased. The probability model used for their parallelized EM algorithm is a Gaussian mixture model. Since it was designed specifically for the CUDA architecture, their parallelized EM algorithm has the best performance when used for CPU-intensive applications. The data parallelism of their EM algorithm was maximized when the clusters were small enough to fit in main memory. The largest data set their parallelized EM algorithm was tested on contained 230,400 data points. Some notable disadvantages of their parallelized EM algorithm were frequent memory conflicts between its threads, and the sensitivity of its performance to the number of threads used.

Cluster validation techniques are discussed in [18], [19]. The clusters produced by a clustering algorithm can be validated by demonstrating the clusters correspond to meaningful patterns in the data set, i.e. it is extremely unlikely the clusters were produced by chance [18]. Cluster validation is necessary to insure the clusters produced by a clustering algorithm are semantically valid. Clusters produced by different clustering algorithms may exhibit different qualities, such as compactness or connectedness. Compact clusters are defined as clusters produced with minimal Total Within-Cluster Variation (TWCV). Connected clusters are defined as clusters produced by grouping together data points in the same neighborhood as each other. KMA searches for the most compact clusters for a data set, while density-based algorithms such as the Density-Based Scan (DBSCAN) algorithm [20] search for the most connected clusters.

Clusters are either validated internally or externally. Techniques for internal validation of clusters include stability validation techniques which measure the consistency of clusters produced by a clustering algorithm which has been applied to several similar data sets. Unlike other techniques for internal validation of clusters, stability validation techniques are not biased towards any particular clustering algorithm.

An important technique for external validation of clusters is testing the statistical significance of the clusters produced by the clustering algorithm [19]. Clusters are tested for statistical significance by showing that the probability of the clustering algorithm producing the clusters by chance does not exceed some predetermined significance level, e.g. a significance level of 5%.

An algorithm for predicting the behavior of automobile drivers is described in [21]. Their predictive algorithm uses computer vision algorithms to: a) detect the automobile's position within its current lane and b) detect changes in the driver's lateral head motion. Furthermore, their predictive algorithm uses data from the automobile's internal sensors to determine its velocity. A Kalman filter is applied to the automobile's reported velocity and its estimated position within its lane to predict its future path. Sparse Bayesian learning is used to predict whether the driver intends to initiate a lane change based on changes in the driver's head motion and the automobile's estimated position within its lane. The reliability of their predictive algorithm was acceptable when tested in scenarios which were similar to the scenarios present in the training data set. However, their predictive algorithm was less reliable in scenarios which differed significantly from those present in the training data set.

Several data mining algorithms for discovering movement patterns of GPS-enabled mobile device users (e.g. smart phone users) are described in [22]. Their algorithm samples the GPS positions of the mobile device users at some predetermined rate and stores those GPS positions in a database. Then, their algorithm normalizes the database by discarding irrelevant data, e.g. errors in the positions reported by the GPS receivers. If the mobile device user stayed near a particular GPS position for some predetermined duration, then that GPS position is considered to be significant. For example, the mobile device user may have stopped at a restaurant for an hour. A movement pattern of a mobile device user is defined as a path which starts and ends at significant GPS positions. These significant GPS positions in a mobile device user's path are correlated to nearby Locations of Interest (LOI), e.g. a department store, that were assumed to have been interesting to the mobile device user.

The support for a movement pattern is defined as the frequency at which the movement pattern occurs in the database. Furthermore, the confidence in the significance of the movement pattern is defined as the relative frequency of the movement pattern in the database with respect to similar movement patterns. If the support and/or confidence for a movement pattern in the database do not exceed certain predetermined thresholds, the movement pattern is considered to not be significant. Any movement patterns discovered in the database which do not meet the minimum criteria for significance (i.e. exceed the minimum thresholds for support and confidence) are excluded from the output of the data mining algorithms.

Their data mining algorithms were tested on 3 artificial data sets constructed by hand. The results from the tests indicate the time complexities of their data mining algorithms grow exponentially as the size of the data sets (i.e. the number of data points) increases. Since their data mining algorithms were only tested with artificial data sets, their accuracy at discovering movement patterns in real data sets could not be verified.

Although research has been done on algorithms for predicting movement patterns by correlating GPS data to other variables [21] and mining movement patterns from 2D GPS-based data sets [22], these algorithms are only capable of predicting or discovering basic movement patterns. The more advanced approach described in this paper efficiently and automatically discovers arbitrarily complex subpaths in ADS-B and FDM data sets using an ant colony algorithm. The subpaths thus discovered can then be data mined to find correlations with other variables, such as an aircraft's altitude and its proximity to the nearest private airport.

Ant colony algorithms have been successfully applied to many combinatorial optimization problems [23], [24], [25], [26], [27]. These algorithms are based on how swarming insects interact with their environment. The Graph-Based Ant System (GBAS) ant colony algorithm uses a construction graph to represent the combinatorial optimization problem to be solved [28]. A construction graph is defined as a special type of directed graph where every path shares a common start node. Time is represented by cycles, which are defined as complete traversals of the construction graph by all the ants (i.e. the agents). The weight assigned to an edge is the probability of an ant traversing it. These weights are calculated from the digital pheromone strength (which evaporates at a rate directly proportional to the number of cycles) and the utility of ants traversing the edge. The utility of ants traversing a particular edge depends on the type of combinatorial optimization problem being solved. The GBAS ant colony algorithm was proved to converge to an optimal solution when certain criteria are met. However, the GBAS ant colony algorithm was not implemented to confirm this convergence proof.

An ant colony algorithm for controlling UA swarms is described in [29]. The current velocity for each UA in the swarm is determined by a vector field. The vectors in the vector field simultaneously direct the UAs in the swarm towards the desired target and away from hazards in the environment. Their ant colony algorithm dynamically constructs a vector field from digital pheromones deposited by UAs in the swarm. Thus, if several UAs in the swarm find the same path to a particular target, other UAs in the swarm will also tend to follow that digital pheromone trail to the target. Similarly, if several UAs find the same path around a particular hazard in the environment, other UAs in the swarm will also tend to follow it. Their ant colony algorithm performs the following steps: 1) the target and hazards in the environment are designated by a human operator at the GCS and 2) the UAs in the swarm autonomously discover an optimal path to the target which avoids the hazards.

An ant colony algorithm for controlling a single UA is described in [30]. Their ant colony algorithm constructs a path the UA can follow to reach a target while avoiding stationary hazards (e.g. ground-based radars) in the environment. Each stationary hazard is assumed to have an associated cost function. Thus, the task of their ant colony algorithm is to find a path that minimizes the cost functions associated with the stationary hazards in the environment. The results from running their ant colony algorithm in several simulated environments indicate their ant colony algorithm can efficiently find a path to a target that avoids stationary hazards in the environment. However, their ant colony algorithm is not applicable to UAs flying in environments with mobile hazards (i.e. dynamic operational environments such as those discussed in [29]).

A preliminary assessment of the risk of a midair collision between a UA and a manned aircraft is discussed in [31]. The likelihood of such a midair collision was estimated using a gas model simulation. The gas model simulation characterized the UA solely by its mass, since aircraft mass is considered a crucial factor in midair collisions. Furthermore, the masses of different types of UAs vary significantly depending on their intended applications. The density of manned aircraft in the gas model simulation was calculated from a single day of data logged by the FAA's Enhanced Traffic Management System (ETMS). ETMS is used by the FAA to archive data from the ground-based radars used by Air Traffic Control (ATC) towers at commercial airports. The UA was considered to have a uniform probability of occupying any position in the airspace, regardless of its mass.

The results from the gas model simulation suggest the risk of midair collision between a UA and a manned aircraft is much higher when the density of manned aircraft is greater, e.g. within FAA-designated airways. The results also suggest smaller UAs (i.e. with less mass) are much less likely to cause severe collision events than larger UAs. Although their preliminary results provide some insight into the total risk of midair collisions between UAs and manned aircraft in the NAS, these findings are not applicable to the risk of midair collisions for specific airspace configurations.

3 Subpath Discovery

The probability of a collision is based on the proximity of the aircraft to each other, the performance characteristics of the aircraft (which are determined by the type of aircraft), and the probabilities of the pilots of those aircraft performing various types of maneuvers. Since pilots flying aircraft near commercial airports are under positive control by Air Traffic Control (ATC), their flight paths are deterministic. Pilots flying aircraft that are not under positive control by ATC (e.g. pilots flying under VFRs in Class E airspace) have nondeterministic flight paths which are much more difficult to predict. Thus, for aircraft flying in Class E airspace, the aircraft's proximity to private airports, rather than commercial airports, is considered more likely to influence the types of maneuvers the pilot performs.

The data mining software we developed, the Aircraft Data Miner (ADM), was used to mine FDM data sets to discover probabilistic models of pilot behavior as a function of the aircraft's type (e.g. a Cessna 172), altitude, and proximity to the nearest private airport. The largest data set we mined with ADM consists of FDM data logged by Garmin G1000 devices onboard aircraft in UND's training fleet. These FDM data were logged during a consecutive 9 month period in 2011.

Most of the data contained in FDM logs for aircraft are not relevant to the analysis of the types of maneuvers performed by pilots during their flights. Also, the FDM data logged by the Garmin G1000 devices are time-ordered sequences of discrete 3D GPS positions occupied by FDM-equipped aircraft, whereas the

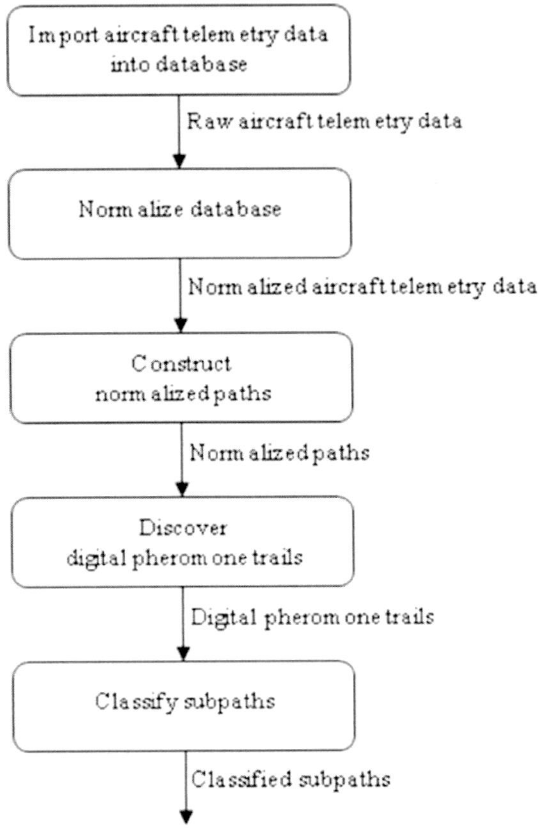

Fig. 2. The flowchart for the five phases of data preprocessing

normalized flight paths of the aircraft (rather than the discrete positions along those flight paths) are more important to the analysis of pilot maneuvers. Thus, it is necessary to extensively preprocess the raw FDM data, which are stored in heterogeneous data formats, into a more useful, homogeneous format prior to data mining.

ADM performs all of its data preprocessing via Structured Query Language (SQL) commands that operate on tables in a relational database (see Fig.2). ADM performs five phases of data preprocessing on the FDM data for each aircraft type in the data set: 1) extracting the FDM data from the log files and inserting the relevant data into a relational database, 2) normalizing the data in the database, 3) constructing normalized flight paths from the discrete 3D GPS positions in the database, 4) discovering digital pheromone trails by finding subpaths which are common to multiple normalized flight paths, and 5) dynamically segmenting the normalized flight paths into subpaths using those digital pheromone trails. These five phases of data preprocessing must be completed prior to data mining.

3.1 Constructing Normalized Paths

The first phase of data preprocessing involves reading the raw FDM data logged by the Garmin G1000 devices, discarding irrelevant data, insuring each aircraft has a unique numeric identifier so it is easily tracked, and inserting the relevant data from each aircraft into the relational database. Data for any flight segments outside of Class E airspace are discarded. During this phase, ADM retains the following fields from the FDM data for each aircraft: a) the latitude, longitude, and altitude, b) the heading, c) the horizontal and vertical velocities, d) the ascent angle, and c) the time of reception. The aircraft's tail number and performance class are specified manually when the FDM data are archived.

The second phase involves normalizing the data in the database. Redundant data, e.g. an aircraft maintaining the same GPS position for several seconds, are discarded. Data for all flights outside of the NAS are also discarded. Since an aircraft's heading is represented as a compass direction in FDM data, it is necessary during preprocessing to normalize the aircraft headings obtained from the FDM data. The headings are normalized with respect to the heading currently considered the straight-flying direction for the aircraft. If the aircraft flew with the same heading *h*for two or more consecutive data points, then *h*would be considered its current straight-flying direction. Thus, to normalize the straight-flying heading h to an angle of 90, it is rotated by *90-h* degrees. Likewise, all other aircraft headings are rotated by the same amount until the next change in the straight-flying direction.

The third phase involves constructing normalized flight paths from the discrete 3D GPS positions for aircraft in the database (i.e. data synthesis). Consecutive data points with the same heading and ascent angles are merged to form the longest possible vectors. The magnitude of these vectors is measured in time, not distance, because while aircraft fly at different speeds (and thus cover different distances in the same amount of time), it is reasonable to assume that the pilots of those aircraft require about the same amount of time to perform the same types of maneuvers.

3.2 Discovering Digital Pheromone Trails

The fourth phase involves the discovery of digital pheromone trails in the normalized flight paths for the aircraft using our ant colony algorithm. In order to discover dynamic classification schemes for the types of maneuvers frequently performed by pilots, each normalized flight path is represented as a separate digital pheromone trail deposited by the aircraft. If every normalized flight path is compared with every other normalized flight path, then subpaths will be discovered that are common to several of the paths. These common subpaths are the areas where digital pheromones from normalized flight paths are reinforcing each other (see Fig.3). For example, if a subpath is common to two normalized flight paths, its corresponding digital pheromone trail will have a strength of 2. Digital pheromone trails with greater strengths are thus more likely to represent actual maneuvers performed by pilots of manned aircraft.

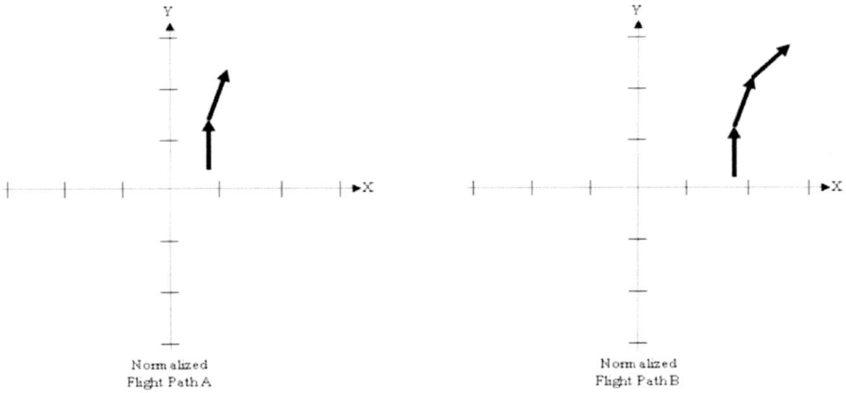

Fig. 3. Normalized paths (projected in 2D) which have one common subpath when D = 1. The common subpath has a strength of 2.

During digital pheromone trail discovery, the shapes of the digital pheromone trails are important, not the GPS positions of their endpoints. Thus, each digital pheromone trail is assigned the same starting point in the coordinate system. Also, normalized flight paths are represented internally as ordered sequences of unit vectors to facilitate comparisons. Every normalized flight path for a given aircraft type is compared with every other normalized flight path for that aircraft type on a vector by vector basis. If two vectors are within some constant maximum diffusion distance *D*of each other, then they are considered a match. Since only the distances between unit vectors will be calculated, *D*must have a value less than 2. Any value for *D*greater than or equal to 2 will result in a case where every vector is within diffusion distance of every other vector. Also, any digital pheromone trails with lengths of less than 5 are discarded, because it is considered unlikely that the pilot of a manned aircraft could complete any significant maneuver in less than 5 seconds.

The algorithm we designed for digital pheromone trail discovery (see Fig.4) performs many distance calculations. To improve its run-time efficiency, a vector proximity map is constructed. A vector proximity map is a 2D Boolean array. It can determine if any two unit vectors are within some maximum diffusion distance *D*of each other in constant run-time. The angles (i.e. heading and ascent angles) for each of the vectors are encoded as integers. These encoded integers are then used as indexes into the vector proximity map. The element in the vector proximity map corresponding to those two vectors is 1 if the vectors are within diffusion distance of each other (i.e. the vectors match), or 0 otherwise.

The strength of a digital pheromone trail is a potential indicator of the frequency at which the pilots performed this type of maneuver. However, all digital pheromone trails discovered during digital pheromone trail discovery are stored in the relational database, irrespective of their strengths. In ant colony algorithms, the strengths of digital pheromone trails decrease over time through evaporation unless they are continually reinforced by new digital pheromones. However, the time elapsed according to the system clock is not a useful control variable for our problem, since FDM data are not necessarily archived at a constant rate. Thus, a better control variable for our problem is the amount of new FDM data that have been archived since the last update to the database.

An incremental algorithm for digital pheromone trail discovery was developed that evaporates existing digital pheromone trails based on the amount of new FDM data being archived. Thus, unless the new FDM data reinforce the existing digital pheromone trails stored in the database, they will eventually evaporate, replaced by new digital pheromone trails. The discovery of digital pheromone trails thus results in a set of frequently occurring subpaths in the FDM data from a specific aircraft type (see Fig.5). Most of the digital pheromone trails in Fig.5 represent maneuvers where the pilots turned to the right at various turn rates. The remaining digital pheromone trails in Fig.5 represent maneuvers where the pilots flew relatively straight or turned to the left at various turn rates. The digital pheromone trails discovered with our algorithm could also result in the chance discovery of unusual and/or counterintuitive maneuvers that pilots of manned aircraft perform.

```
Algorithm Discover Digital Pheromone Trails.
  Input: Set of N normalized flight paths.
  Output: Set of digital pheromone trails.

  D ← maximum diffusion distance
  for p = 1 to N do
    for each vector u in path p do
      for q = p to N do
        if length(p) = length(q) then
          t ← new pheromone trail
          for each vector v in path q do
            if distance(u,v) > D then
              if length(t) > 0 then
                if t is not in database then
                  Insert t into database.
                t ← new pheromone trail
            else
              Append u to t.
```

Fig. 4. The pseudocode for the digital pheromone trail discovery algorithm

3.3 Classifying Subpaths

The fifth and final phase of data preprocessing involves segmenting the normalized flight paths into subpaths using the digital pheromone trails discovered by

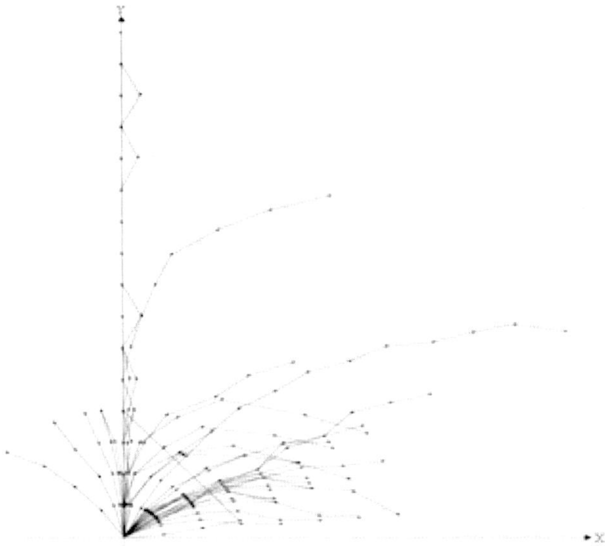

Fig. 5. The digital pheromone trails (projected in 2D) used to classify the subpaths. Only digital pheromone trails with strengths of at least 5,000 are shown.

the previous phase. The subpaths identified during this phase are stored in the database for later retrieval during data mining. Each normalized flight path $P$is segmented into subpaths in an iterative manner, starting with its first vector. The current vector $v_p$from $P$is compared with the first vector $v_t$from every digital pheromone trail in the database whose length is less than or equal to the length of P. If the vectors $v_p$and $v_t$are within some maximum diffusion distance $D$of each other, then the digital pheromone trail is a potential match. If the vectors $v_p$and v_t are not within some maximum diffusion distance $D$of each other, then the digital pheromone trail is not a potential match.

Next, each of the potentially matching digital pheromone trails is compared with every subpath in $P$of the same length which has the same starting vector. This determines if any subpath of Pcompletely matches one of the digital pheromone trails. From the set of complete matches to subpaths in P, the digital pheromone trail with the greatest strength is used to classify the subpath. If there are multiple such digital pheromone trails, the longest digital pheromone trail from the set of strongest matches is used to classify the subpath. The matching digital pheromone trail, having some length l_t, will thus match some subpath from $P$of lengthl_t. The next subpath segmented from P will start immediately after the end of the previous subpath. If there are only a few normalized flight paths for a particular aircraft type, it is possible for a normalized flight path to have subpaths that are not common to any other normalized flight paths. Such subpaths are defined by exclusion and cannot be classified.

The set of digital pheromone trails discovered for a particular aircraft type thus form a dynamic classification scheme for the types of maneuvers a pilot is likely to perform when flying that type of aircraft. Each digital pheromone trail is a potentially unique type of maneuver that was intentionally performed by pilots. In some cases, there are may be two or more digital pheromone trails with different strengths that match the same subpath in a normalized flight path. In these cases the digital pheromone trail with the greatest strength (i.e. the one that occurs most frequently in the data) is always used. However, since we do not know the actual intentions of the pilot, it is possible (though unlikely) that the pilot actually performed a maneuver corresponding to a digital pheromone trail of lesser strength.

There may have been other factors affecting the aircraft's heading, ascent angle, and altitude than just the pilot's control inputs. Also, the uncertainties in the aircraft's horizontal and vertical position (inherent to GPS-based FDM data logging devices) introduce a measurable amount of error. To correct for these anomalies, ADM rounds up the aircraft's heading and ascent angles to the nearest multiple of 2, and the aircraft's altitude to the nearest multiple of 2 meters. Thus, the values for the heading, ascent angle, and altitude of the aircraft are considered to be accurate indicators of the pilot's intent to within 1 for headings and ascent angles, and to within 1 meter for altitudes. This is a valid assumption in most cases, since, for example, the standard turn rate for a pilot is 2 per second, which means pilots will not normally turn their aircraft at a rate of less than 2 per second.

The aircraft's altitude, proximity to the nearest commercial airport, and proximity to the nearest private airport at the start of each subpath are calculated and/or stored in the database for later retrieval. The proximity to the nearest commercial or private airport is the geodesic distance from the aircraft's latitude/longitude position to the latitude/longitude position of the nearest commercial or private airport, respectively. Proximity to airports is calculated by searching an airport database for commercial or private airports, respectively, which are near the aircraft, finding the distance from the aircraft to each of the nearby commercial or private airports, respectively, and then selecting the minimum of those distances.

If a high degree of accuracy is desired for the distances calculated between two points specified as latitudes and longitudes, Vincenty's inverse method [32] is preferred, since the results from Vincenty's inverse method are accurate to within half a millimeter. Although very accurate, algorithms based on Vincenty's inverse method are computationally intensive. Thus, our algorithm for calculating the aircraft's proximity to an airport was designed to reduce the total number of distance calculations performed with Vincenty's inverse method.

Our algorithm represents the contiguous area of the NAS as a large grid of cells. A cell in the grid measures 50 meters on each side. Since the curvature of the Earth over a 50 square meters area is negligible, any curvature within the cells can be ignored. The number of cells in every column in the grid is 55,121 cells (which is equivalent to 2,756,050 meters). The number of cells in the rows of the grid varies from 82,259 cells (or 4,112,950 meters) to 117,052 cells (or 5,852,600 meters), depending on the row's latitude. To map a point specified as latitude and longitude to a point within the grid, our algorithm only needs to calculate two distances using Vincenty's inverse method—from the western edge of the grid to the point and from the southern edge of the grid to the point.

After calculating those two distances, our algorithm divides both distances by the length of a side of a grid cell (i.e. 50 meters), rounding down to the nearest integer, to obtain the x and y coordinates for the point within the grid. The grid positions for all the airports (both private and commercial) in the database are calculated prior to the subpath classification phase, and stored for later retrieval. The airport's type (i.e. either private or commercial) is stored along with its grid position. Then, during subpath classification, each latitude/longitude point along the aircraft's path is mapped to its corresponding point within the grid.

Since the grid positions for all the airports in the database are calculated offline, our algorithm can efficiently determine which airports of a specific type are near the aircraft using the aircraft's current position within the grid. First, our algorithm searches for airports of that specific type within the same cell as the aircraft. If there are not any airports of that specific type in the same cell as the aircraft, our algorithm searches all the cells that border that cell, and so on, until cells containing one or more airports of that specific type are found. Once our algorithm finds the cell(s) containing the airports of that specific type which are closest to the aircraft's cell, it calculates the distance between the aircraft and each of those airports, and uses the minimum for the aircraft's proximity measurement.

Although the volume of controlled airspace around commercial airports varies from airport to airport, the maximum volume of controlled airspace around any airport in the NAS is 27,780 meters horizontally and 3,000 meters vertically. Thus, any data points from aircraft within this maximum volume of controlled airspace around some airport were considered to be within controlled airspace. These data points were therefore excluded from further consideration.

4 Data Mining Subpaths

ADM performs all of its data mining via SQL commands that operate on tables in a relational database. ADM performs its data mining in two phases (see Fig.6): 1) altitude mining and 2) proximity mining. These two data mining phases are performed separately for the data from each aircraft type in the FDM data set. The phases must occur in sequence to produce correct results. Both of the data mining phases operate on the same relational database. This relational database stores results from each of the data mining phases. Data are grouped into tables based on their respective aircraft types.

ADM uses a database-oriented implementation of GKA for altitude mining. Either GKA or a database-oriented hybrid algorithm, which combines the EM

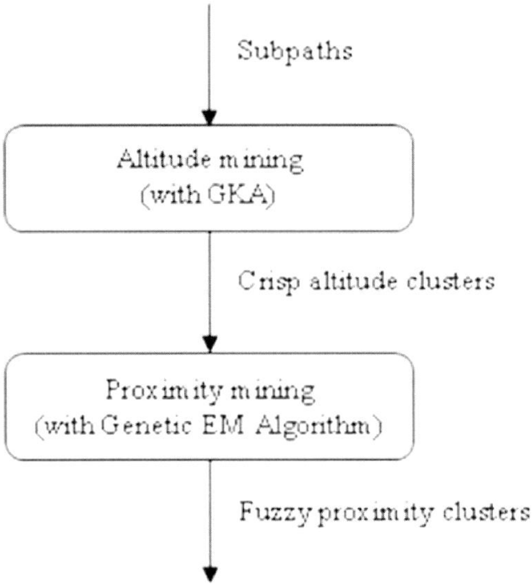

Fig. 6. The flowchart for the two phases of data mining

algorithm with a genetic algorithm, can be used for the proximity mining phase. The genetic EM algorithm (see Fig.7) uses a Gaussian mixture model for the probability distributions. ADM data mines the data for each aircraft type as follows. In the first phase, the subpaths (i.e. the data points in the clusters) discovered in the data for the aircraft type are automatically clustered into $k_1$clusters with respect to the altitude at the start of each subpath (see Fig.8). In the second phase, the best solution found during the first phase is data mined further. Each of the k_1 clusters from that solution is data mined with respect to the aircraft's proximity to the nearest private airport at the start of each subpath (see Fig.9). This subdivides each of the $k_1$altitude clusters into $k_2$proximity clusters, resulting in a cluster hierarchy.

ADM has both sequential and parallel algorithms for data preprocessing and data mining. ADM's parallel algorithms were implemented using the Parallel Virtual Machine (PVM) system [33] for parallel computing. In order to prevent key conflicts during parallel execution, temporary keys are used during the normalized flight path construction and digital pheromone trail discovery phases of data preprocessing, as well as both of the data mining phases. Periodic communication between the master and slave processes was necessary in order to balance the load on the relational database server. However, if a distributed file system or distributed database server is used, periodic communication between the master and slave processes should not be necessary.

Algorithm Genetic Expectation-Maximization.
Input: Set of N subpaths.
Output: Set of N subpaths partitioned into K clusters.

$P \leftarrow M$ clusterings with random parameters
$G \leftarrow$ number of generations
$P_m \leftarrow$ mutation probability
$X_i \leftarrow$ ith data point
$H_{ij} \leftarrow$ membership probability of X_i for jth cluster
$\mu_j \leftarrow$ mean value for jth cluster
$\sigma_j \leftarrow$ standard deviation for jth cluster
$\pi_j \leftarrow$ weight for jth cluster
$\eta_{ij} \leftarrow$ Gaussian for X_i in jth cluster
for $t = 1$ **to** G **do**
for each candidate solution S_t **in** P **do**

$$\eta_{ij} \leftarrow \frac{1}{\sqrt{2\pi\sigma_j^2}} e^{\frac{-(X_i-\mu_j)^2}{2\sigma_j^2}}$$

$$H_{ij} \leftarrow \frac{\pi_j \cdot \eta_{ij}}{\sum_{r=1}^{K} \pi_r \cdot \eta_{ir}}$$

$$\mu_j \leftarrow \frac{\sum_{s=1}^{N} H_{sj} \cdot X_s}{\sum_{s=1}^{N} H_{sj}}$$

$$\sigma_j \leftarrow \frac{\sum_{s=1}^{N} H_{sj}(X_s - \mu_j)^2}{\sum_{s=1}^{N} H_{sj}}$$

$$\pi_j \leftarrow \frac{\sum_{s=1}^{N} H_{sj}}{\sum_{r=1}^{K}\sum_{s=1}^{N} H_{sr}}$$

for each candidate solution S_t **in** P **do**
$r \leftarrow$ random number between 0 and 1
if $r < P_m$ **then**
$S_{t+1} \leftarrow$ randomly perturbed S_t
else
$S_{t+1} \leftarrow S_t$
return candidate solution with minimal total variance.

Fig. 7. The pseudocode for the genetic EM algorithm

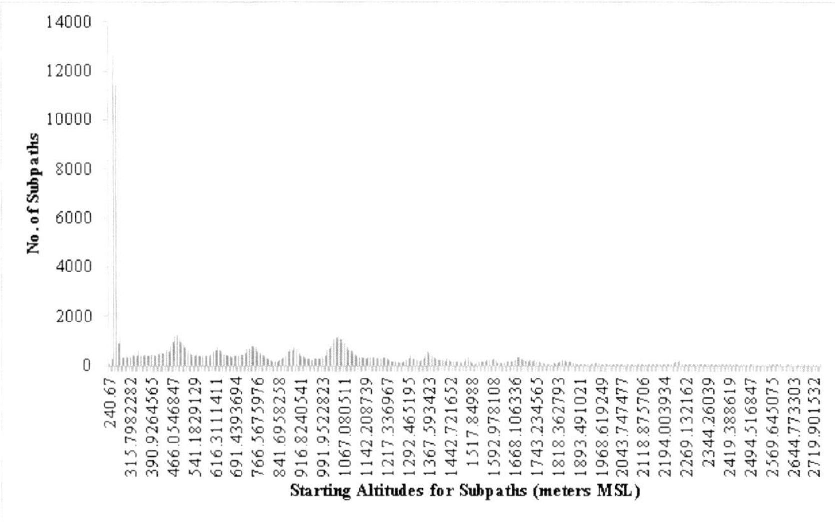

Fig. 8. The histogram for discovered subpaths with respect to starting altitudes

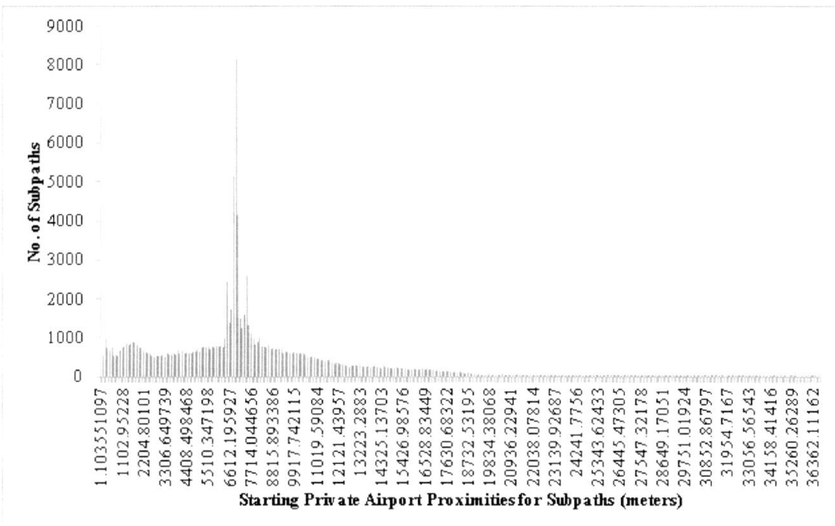

Fig. 9. The histogram for discovered subpaths with respect to starting proximities to private airports

5 Results

ADM sequentially preprocessed the raw FDM data set, which consists of approximately 104 gigabytes of data, to find subpaths in the normalized flight paths of aircraft. The raw FDM data set was sequentially preprocessed into 111,458 subpaths using 21 discovered digital pheromone trails with strengths of 5,000 or greater. Both the subpaths and the digital pheromone trails were stored in tables in a relational database. After sequentially preprocessing the raw FDM data set, ADM performed parallel data mining of the resultant subpaths on a Beowulf cluster with the Network File System (NFS).

ADM evolved 6 candidate solutions for 200 generations during both phases of data mining. To test ADM's parallel data mining algorithms, the data mining was performed on 3 nodes of a Beowulf cluster. ADM's master process did not perform any data mining itself. Instead, the master process merely divided the task and spawned slave processes on the remaining nodes in the Beowulf cluster. Thus, each of the 2 slave nodes was assigned 3 candidate solutions for parallel data mining. The data mining results are shown in Tables 1 and 2. These results have a statistical significance level of 0.05% when Pearson's chi-square test is used.

ADM first categorized the data points into one of 2 clusters based on the aircraft's altitude. This resulted in a cluster of low altitudes and a cluster of high altitudes. Then, for each of these clusters, ADM categorized the data points within the cluster into one of 2 subclusters based on the aircraft's proximity to the nearest private airport. Since a genetic EM algorithm was used for proximity mining, the proximity clusters produced are not crisp clusters. The membership probabilities for digital pheromone trails in the first altitude subcluster (see

Table 1. The histogram for discovered subpaths with respect to starting proximities to private airports

Pheromone Trail ID	Frequency	Probability of Membership in Cluster 1	Probability of Membership in Cluster 2
115	1	0%	100%
264	2	0%	100%
367	6	0%	100%
236	9	0%	100%
366	9	0%	100%
250	16	0%	100%
232	18	0%	100%
2131	22	0%	100%
396	27	0%	100%
214	31	0.45%	99.55%
379	32	1.25%	98.75%
246	49	1.27%	98.73%
248	199	0.8%	99.2%
216	272	0.31%	99.69%
19	380	1.17%	98.83%
241	570	0.2%	99.8%
226	616	0.25%	99.75%
30	655	0.49%	99.51%
213	60,419	1.3%	98.7%

Table 1) show a strong bias towards only one of the possible proximity clusters. This suggests that, at relatively low altitudes, the maneuvers naturally form into only one proximity cluster. However, the membership probabilities for the second altitude cluster (see Table 2) do not show a strong bias towards either of the possible proximity clusters. This suggests that, at higher altitudes, pilot maneuvers naturally group into two proximity clusters—a cluster of aircraft near to a private airport and a cluster of aircraft far from a private airport. Thus, a pilot flying an aircraft at lower altitudes would be unlikely to be influenced by the aircraft's proximity to nearby private airports. At higher altitudes, however, the pilot would be more likely to perform different types of maneuvers based on whether the aircraft is near to or far from a private airport.

1. The frequencies and membership probabilities for the maneuvers discovered in the first altitude cluster (with altitudes from 240.67 to 864.02 meters MSL) for private aircraft.
2. The frequencies and membership probabilities for the maneuvers discovered in the second altitude cluster (with altitudes from 864.05 meters to 2,742.44 meters MSL) for private aircraft.

Table 2.

Pheromone Trail ID	Frequency	Probability of Membership in Cluster 1	Probability of Membership in Cluster 2
146	1	100%	0%
227	1	62%	38%
264	1	80%	20%
379	1	52%	48%
367	2	79.5%	20.5%
396	4	57.5%	42.5%
232	5	73%	27%
366	7	67.14%	32.86%
236	8	75%	25%
250	8	74%	26%
214	10	74.2%	25.8%
2131	12	74.75%	25.25%
246	17	68.09%	31.91%
248	73	69.53%	30.47%
216	98	69.63%	30.37%
19	252	67.19%	32.81%
226	286	67.99%	32.01%
241	287	68.09%	31.91%
30	491	67.77%	32.23%
213	42,858	69.45%	30.55%

6 Conclusion

Probabilistic models for the behavior of pilots of GA aircraft flying in Class E airspace were obtained by data mining a large FDM data set. This data set consisted of FDM data from a consecutive 9 month period in 2011. First, the raw FDM data set was sequentially preprocessed into 111,458 subpaths. Then, these subpaths were data mined in parallel with respect to 1) the starting altitude of the subpaths (using GKA) and 2) the starting private proximities of the subpaths (using the genetic EM algorithm). The data mining results have a statistical significance level of 0.05% when Pearson's chi-square test is used.

Since mining a large data set is a lengthy process, ADM was modified to execute its algorithms in parallel on a Beowulf cluster. Future work could include implementing crossover between cluster nodes for the parallelized GKA and testing the parallel data mining and data preprocessing algorithms on a Beowulf cluster with a distributed file system or a distributed database server.

Each digital pheromone trail discovered by our ant colony algorithm is a potentially unique maneuver performed by the pilots of manned aircraft flying under VFRs in Class E airspace. Thus, our ant colony algorithm could also result in the chance discovery of unusual maneuvers such pilots may sometimes perform. The RM subsystem of GPAR-RMS was intended for the chance discovery of the probability of a midair collision between the UA and other proximate aircraft in the airspace. The chance discovery of unusual maneuvers could also be important to collision avoidance for the RM subsystem of any ground-based sense-and-avoid system.

Acknowledgments. Some of this work was performed with funding from the U.S. Air Force (contract #FA4861-06-C-C006), the FAA (grant #7-C-GA-UND), and the North Dakota EPSCoR program (NSF grant #EPS-0814442).

References

1. Dalamagkidis, K., Valavanis, K.P., Piegl, L.A.: On Unmanned Aircraft Systems Issues, Challenges, and Operational Restrictions Preventing Integration into the National Airspace System. Progress in Aerospace Sciences 44, 503–519 (2008)
2. Federal Aviation Administration, `http://www.faa.gov/library/manuals/aviation/risk_management/ss_handbook`
3. Marsh, R., Ogaard, K., Kary, M., Nordlie, J., Theisen, C.: Development of a Mobile Information Display System for UAS Operations in North Dakota. International Journal of Computer Information Systems and Industrial Management Applications 3, 435–443 (2011)
4. Federal Aviation Administration, `http://ecfr.gpoaccess.gov/cgi/t/text/text-idx?c=ecfr&tpl=/ecfrbrowse/Title14/14tab_02.tpl`
5. Marsh, R., Ogaard, K.: Mining Heterogeneous ADS-B Data Sets for Probabilistic Models of Pilot Behavior. In: Proceedings of the 10th IEEE International Conference on Data Mining Workshops, pp. 606–612. IEEE Press, New York (2010)
6. Duda, R.O., Hart, P.E., Stork, D.G.: Pattern Classification. John Wiley and Sons, New York (2001)
7. Krishna, K., Murty, M.N.: Genetic K-Means Algorithm. IEEE Transactions on Systems, Man, and Cybernetics 29, 433–439 (1999)

8. Rudolph, G.: Convergence Analysis of Canonical Genetic Algorithms. IEEE Transactions on Neural Networks 5, 96–101 (1994)
9. Eiben, A.E., Aarts, E.H.L., Van Hee, K.M.: Global Convergence of Genetic Algorithms: A Markov Chain Analysis. In: Schwefel, H.-P., Männer, R. (eds.) PPSN 1990. LNCS, vol. 496, pp. 3–12. Springer, Heidelberg (1991)
10. Lu, Y., Lu, S., Fotouhi, F., Deng, Y., Brown, S.J.: Incremental Genetic K-Means Algorithm and its Application in Gene Expression Data Analysis. BMC Bioinformatics 5 (2004)
11. Al-Shboul, B., Myaeng, S.H.: Initializing K-Means using Genetic Algorithms. World Academy of Science, Engineering, and Technology 54, 114–118 (2009)
12. Chandar, K., Kumar, D., Kumar, V.: Enhancing Cluster Compactness using Genetic Algorithm Initialized K-Means. International Journal of Software Engineering Research and Practices 1, 20–24 (2011)
13. Kumar, N.S.L.P., Satoor, S., Buck, I.: Fast Parallel Expectation Maximization for Gaussian Mixture Models on GPUs using CUDA. In: Proceedings of the 11th IEEE International Conference on High Performance Computing and Communications, pp. 103–109. IEEE Press, New York (2009)
14. Dempster, A.P., Laird, N.M., Rubin, D.B.: Maximum Likelihood from Incomplete Data via the EM Algorithm. Journal of the Royal Statistical Society 39, 1–38 (1977)
15. Manning, C.D., Schütze, H.: Foundations of Statistical Natural Language Processing. The MIT Press, Cambridge (1999)
16. Moore, A.: Very Fast EM-Based Mixture Model Clustering using Multiresolution kd-Trees. In: Proceedings of the 11th Conference on Advances in Neural Information Processing Systems, pp. 543–549. The MIT Press, Cambridge (1998)
17. Plant, C., Böhm, C.: Parallel EM-Clustering: Fast Convergence by Asynchronous Model Updates. In: Proceedings of the 10th IEEE International Conference on Data Mining Workshops, pp. 178–185. IEEE Press, New York (2010)
18. Handl, J., Knowles, J., Kell, D.B.: Computational Cluster Validation in Post-Genomic Data Analysis. Bioinformatics 21, 3201–3212 (2005)
19. Halkidi, M., Batistakis, Y., Vazirgiannis, M.: On Clustering Validation Techniques. Journal of Intelligent Information Systems 17, 107–145 (2001)
20. Ester, M., Kriegel, H.P., Sander, J., Xu, X.: A Density-Based Algorithm for Discovering Clusters in Large Spatial Databases with Noise. In: Proceedings of the 2nd International Conference on Knowledge Discovery and Data Mining. AAAI Press, Menlo Park (1996)
21. McCall, J.C., Wipf, D.P., Trivedi, M.M., Rao, B.D.: Lane Change Intent Analysis using Robust Operators and Sparse Bayesian Learning. IEEE Transactions on Intelligent Transportation Systems 8, 431–440 (2007)
22. Taniar, D., Goh, J.: On Mining Movement Patterns from Mobile Users. International Journal of Distributed Sensor Networks 3, 69–86 (2007)
23. Ando, Y., Fukazawa, Y., Masutani, O., Iwasaki, H., Honiden, S.: Performance of Pheromone Model for Predicting Traffic Congestion. In: Proceedings of the 5th International Joint Conference on Autonomous Agents and Multiagent Systems, pp. 73–80. ACM Press, New York (2006)
24. Chu, H.N., Glad, A., Simonin, O., Sempé, F., Drogoul, A., Charpillet, F.: Swarm Approaches for the Patrolling Problem, Information Propagation vs. Pheromone Evaporation. In: Proceedings of the 19th IEEE International Conference on Tools with Artificial Intelligence, pp. 442–449. IEEE Press, New York (2007)

25. Narzt, W., Pomberger, G., Wilflingseder, U., Seimel, O., Kolb, D., Wieghardt, J., Hörtner, H., Haring, R.: Self-Organization in Traffic Networks by Digital Pheromones. In: Proceedings of the 10th IEEE Intelligent Transportation Systems Conference, pp. 490–495. IEEE Press, New York (2007)
26. Kalivarapu, V., Foo, J.L., Winer, E.: Improving Solution Characteristics of Particle Swarm Optimization using Digital Pheromones. Journal of Structural and Multidisciplinary Optimization 37, 415–427 (2008)
27. Kannampallil, T.G., Fu, W.-T.: Trail Patterns in Social Tagging Systems: Role of Tags as Digital Pheromones. In: Schmorrow, D.D., Estabrooke, I.V., Grootjen, M. (eds.) FAC 2009. LNCS, vol. 5638, pp. 165–174. Springer, Heidelberg (2009)
28. Gutjahr, W.J.: A Graph-Based Ant System and its Convergence. Future Generation Computer Systems 16, 873–888 (2000)
29. Parunak, H.V.D., Purcell, M., O'Connell, R.: Digital Pheromones for Autonomous Coordination of Swarming UAVs. In: Proceedings of the 1st AIAA Technical Conference and Workshop on Unmanned Aerospace Vehicles, Systems, Technologies, and Operations. AIAA Press, Reston (2002)
30. Ma, G., Duan, H., Liu, S.: Improved Ant Colony Algorithm for Global Optimal Trajectory Planning of UAV under Complex Environment. International Journal of Computer Science and Applications 4, 57–68 (2007)
31. Weibel, R.E., Hansman Jr., R.J.: Safety Considerations for Operation of Different Classes of UAVs in the NAS. In: Proceedings of the 3rd AIAA Unmanned Unlimited Technical Conference, Workshop, and Exhibit. AIAA Press, Reston (2004)
32. Vincenty, T.: Direct and Inverse Solutions of Geodesics on the Ellipsoid with Application of Nested Equations. Survey Review 23, 88–93 (1975)
33. Geist, A., Bequelin, A., Dongarra, J., Jiang, W., Manchek, R., Sunderam, V.S.: PVM: Parallel Virtual Machine: A User's Guide and Tutorial for Network Parallel Computing. The MIT Press, Cambridge (1994)

Constructing Feature Set by Using Temporal Clustering of Term Usages in Document Categorization

Hidenao Abe[1] and Shusaku Tsumoto[2]

[1] Faculty of Information and Communications, Bunkyo University 1100 Namegaya, Chigasaki, Kanagawa, 2538550, Japan
hidenao@shonan.bunkyo.ac.jp

[2] Department of Medical Informatics, Shimane University, School of Medicine 89-1 Enya-cho, Izumo, Shimane 693-8501, Japan
abe@med.shimane-u.ac.jp, tsumoto@computer.org

Abstract. For discovering some chances in documents with temporal context, it is important to handle their contents represented as words and phrases, called "keywords". However, in conventional methods, keywords are selected based on their frequency and/or a particular importance index such as tf-idf throughout their observed period. In this chapter, we describe a method for characterizing large number of documents, considering the temporal features of appeared terms, by obtaining document clusters based on the similarities between the document that are characterized by the temporal patterns of an importance index for considering temporal differences in term usages. As an experiment, we performed document clustering for four sets of bibliographical documents using two feature sets: popular feature terms appearances and the appearances of temporal patterns for each document. Then, we compared the time dependencies of the two document clustering results. Our feature construction method succeeded in representing the time differences in the documents using features based on temporal patterns.

Keywords: Text Mining, Temporal Clustering, Text Categorization, Feature Construction.

1 Introduction

The amount and ratio of electrically stored information, including documents, have been increasing every year. The amount of such data often overwhelms people who want to discover valuable information in the stored data. There is great needs for the discovery of many kinds of chances [13], which is valuable information for decision-making, in the massive amount of data. It is important to determine the features representing temporal information of each document by focusing on the chance for distinguishing temporal context of the documents, such as periodical moods and their basics.

By using the features representing temporal information of each document, people can discover valuable information leading to chance from the stored documents. There are two major approaches for extracting the information: one is based on the documents, and the other is based on words and phrases[1] included in the documents. The former approach includes document categorization [18] and document classification. In this approach, categories are defined by clustering, probabilistic distributions, and manual

[1] We refere to "words and phrases" as "terms" in this article.

Y. Ohsawa & A. Abe (Eds.): Advances in Chance Discovery, SCI 423, pp. 215–229.
springerlink.com

labeling by human experts. The later approach focuses on appearance and its remarkable trends such as phrase clustering [9] and topic modeling [10]. This remarkable trend detection has attracted attentions in data mining and knowledge discovery fields such as in emergent term detection (ETD) methods [7,6].

In the former approach for finding temporal context in the documents, the selection of adequate representative keywords from all of the terms appearing in document categorization and classification is difficult, as pointed out in [17]. In addition, the conventional approach does not consider temporal differences in the document categories, as described in [14]. From the viewpoint of the conventional datamining methodology, this approach for representing temporal context features can be considered to be a novel way of constructing such features from any kind of temporally published documents.

For the abovementioned issues, we focus on the temporal behaviors of importance indices of terms and their temporal patterns. As for basic considerations, we should treat the index values of the terms for each periodical set of documents, and should group similar temporally contextual documents based on the representative terms. The identification of representative terms by using temporal patterns enables reduction in the total number of representative terms. A method of combining temporal pattern extractions for detection of representative terms for the document clustering was developed [2].

In this chapter, we perform a comparison of the document clustering results between the result of the representative terms extracted from the temporal patterns and those determined using conventional representative term selection. In Section 2, we describe the document clustering method with considering temporal patterns of term usages represented as the temporal similarities of the importance index. In this experiment, the two datasets consisting of the feature sets using the titles of the four datamining related conferences ware constructed: popular feature terms appearances and the temporal pattern appearances for each document. Then, in Section 3, we performed document clustering for four sets of bibliographical documents by using the two feature sets to compare the time dependencies of the two document clustering results. With regard to the results, we discuss the efficiency of representing temporal information using term appearances included the temporal patterns as a feature set of temporally published documents. In Section 4, we summarize this chapter.

2 Method for Categorizing Documents Considering Time Differences in Term Usage Patterns

In this method, we assume that given sets of documents are published in each period; annually, monthly, daily, hourly, and so forth. Thus, the sets of documents consist of each set of documents published in particular period. At the time of categorizing the documents, we know all of the documents that are published in the past. The task is to find document clusters C_K for an integer K containing the documents published in both uniform periods and different periods with the same temporal context based on the features of the documents.

In order to categorize given whole sets of documents, taking into consideration the differences in the contents that are influenced by time differences, a method to group the documents with a clustering algorithm and perform the feature construction using a temporal pattern extraction process is proposed.

In conventional document categorization methods, the feature terms are determined using entire documents and a particular importance index. The documents are featured by the appearances of highly ranked feature terms throughout all of the periods. Fig. 1 illustrates the typical process of conventional document categorization methods.

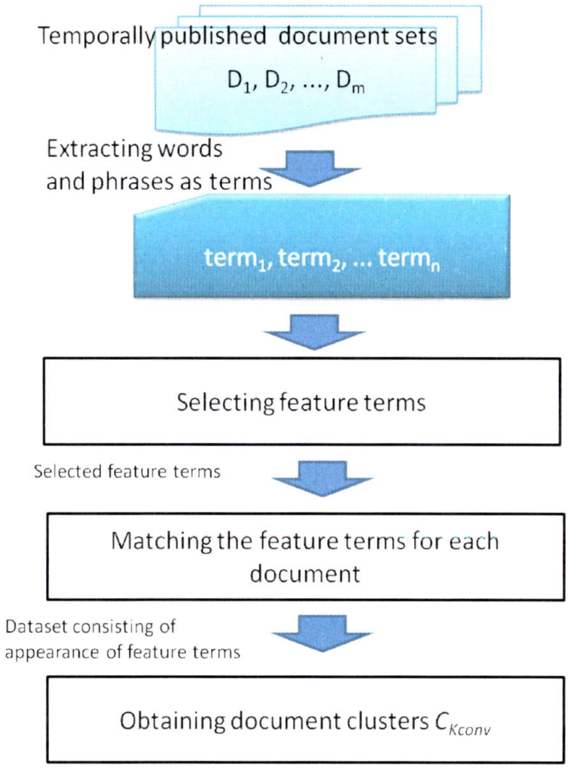

Fig. 1. Typical document categorization process in conventional methods

Feature set construction using temporal pattern extraction was introduced to handle time differences of term usages in each period. The term appearances are converted to matching information for the temporal patterns for characterizing each document. The process of our proposal is illustrated in Fig. 2.

2.1 Grouping Terms by Using Temporal Importance Index Values

In this section, we describe the following two steps for extracting temporal patterns of the terms of each importance index:

1. Importance index calculation
2. Temporal pattern extraction

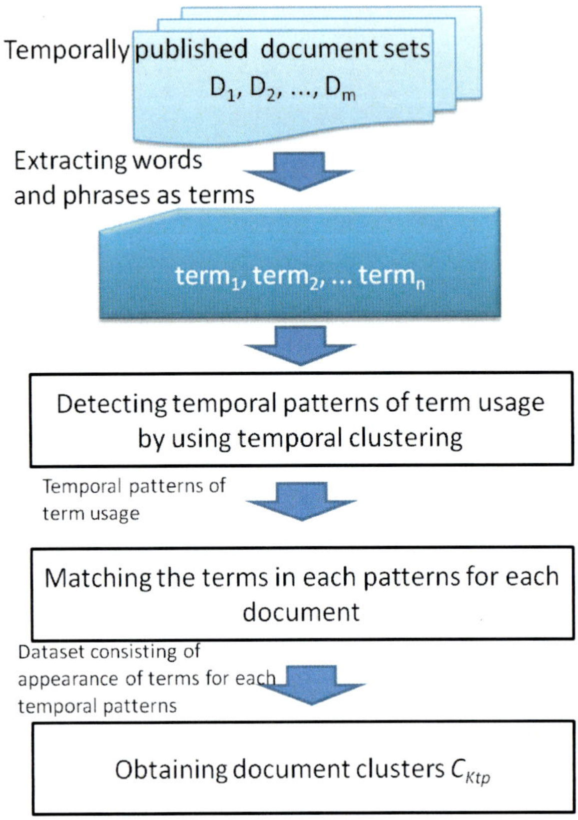

Fig. 2. Overview of our feature set construction method for document categorization

There are some conventional methods of extracting technical terms in a corpus on the basis of each particular importance index [6]. Although these methods calculate each to extract technical terms, information on the importance of each term is lost by cutting off the information with a threshold value. We suggest separating term determination and temporal trend detection based on importance indices. By separating these phases, we can calculate different types of importance indices to obtain a dataset consisting of the values of these indices for each term. Subsequently, we can apply many types of temporal analysis methods to the dataset based on statistical analysis, clustering, and machine learning algorithms.

After determining terms in the given corpus, the system calculates multiple term importance indices for the set of documents in each period. Based on the frequencies of the appearance of each terms in the set of documents, we can calculate well-known importance indices such as simple document frequency, appearance ratio, and term frequency divided by inversed document frequency (tf-idf) [15]. Further, in the proposed method, we can assume the degrees of co-occurrence such as Jaccard's matching coefficient [3] and the χ^2 statistics for terms consisting of multiple words to be the importance indices. Because their values are influenced by the usage of the terms in the given corpus, we can observe the differences in term usages among different periods.

In the proposed method, we suggest treating these indices explicitly as a temporal dataset. The features of this dataset consist of the values of prepared indices for each period. Fig. 3 shows an example of the dataset consisting of an importance index for each period. The value v_{ij} is calculated for a term $term_i$ in the document set D_{t_j} of the period t_j. Then, the framework provides the choice of some adequate trend extraction methods to the dataset. By applying an adequate time-series analysis method, users can discover temporal patterns by processing the values in the rows in Fig. 3.

	t_1	t_2	...	t_n
$term_1$	v_{11}	v_{12}	... v_{1j} ...	v_{1n}
$term_2$	v_{21}	v_{22}	... v_{2j} ...	v_{2n}
	v_{i1} ⋮	v_{i2} ⋮	... v_{ij} ...	v_{in} ⋮
$term_m$	v_{m1}	v_{m2}	... v_{mj} ...	v_{mn}

Fig. 3. Example of a dataset consisting of an importance index

The outputs are the groups of the terms, c_k consisting of terms with similar temporal behavior on the importance index throughout the periods from t_1 to t_m.

2.2 Obtaining Datasets of the Documents Based on Vector Space Model

Vector space model such as bag of words/terms are very popular for characterizing the documents as a dataset for identifying text categories. In this proposal, we use a vector space model for characterizing each document d_x using both the feature terms and the temporal patterns of term usages.

In the conventional document clustering method shown in Fig. 1, following two processes are used to obtain the dataset to make document clusters. The system selects feature terms from the given terms to characterize each document. The, it convert the documents into a vector representation by matching the feature terms and the terms included in the documents. The process for obtaining an instance $\boldsymbol{x}_{conv}$ for each document d_x is shown in Fig. 4.

The process for obtaining instances, $\boldsymbol{x}_{tp}$, for clustering the given document using temporal patterns is shown in Fig. 5. The system matches the terms included in a temporal pattern c_k for a document d_x. It iterates this process for each temporal pattern. The features consist of simple appearance information or weighted information about the terms included in c_k. The number of features is equal to the number of temporal patterns $|c_k|$.

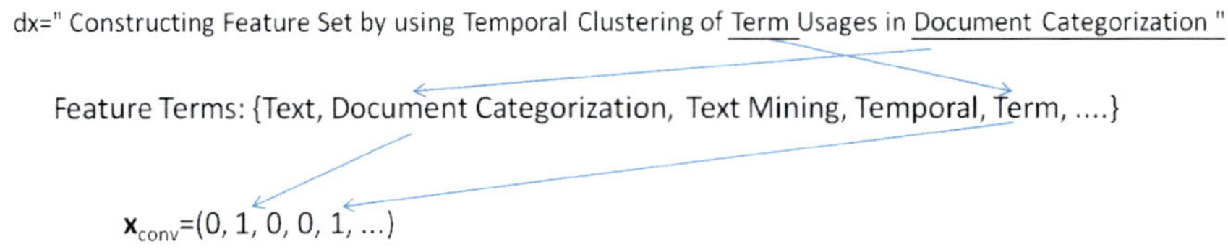

Fig. 4. Process for obtaining an instance consisting of the Boolean vector based on feature terms

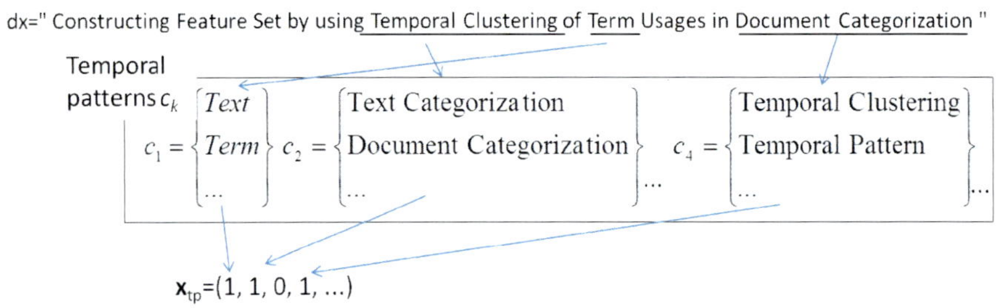

Fig. 5. Process for obtaining an instance consisting of the Boolean vector based on the temporal patterns of terms' usage

Thus, the output of our method is similar to the output of the text categorization using co-clustering approach. However, most conventional methods using the co-clustering approach [4] did not consider the time dependency of the term usage differences.

3 Experiment: Comparing Time Dependency of Document Clustering Results Using Two Feature Sets

In this experiment, we show the results of document clustering by using temporal patterns of an importance index in text mining are shown, as described in Section 2. The annual sets of the titles of the following four data mining related conferences[2]; KDD, PKDD, PAKDD, and ICDM; as used as the input of temporal document sets.

We use the titles of the four conferences as temporal sets of documents. The descriptions of the sets of the documents are shown in Table 1.

For the sets of documents, we assume each article to be one document d_x. Hence, the set of document D_{t_j} in each period t_j consists of the document $d_x \in D_{t_j}$. The size of the set of D_{t_j} is described as $|D_{t_j}|$. For example, the size D_{2008} of KDD, $|D_{2008}|$ is 134.

There are some well-known importance indices of words and phrases in a corpus. Term frequency divided by inversed document frequency (tf-idf) is one of the indices commonly used for measuring the importance of the terms. Representative terms based on the tf-idf values for the entire sets of documents for each conference were selected

[2] These titles are the part of the collection by DBLP [1].

Table 1. Description of the numbers of titles

Year	KDD	PKDD	PAKDD	ICDM
1994	40			
1995	58			
1996	74			
1997	65	43		
1998	68	56	51	
1999	93	82	72	
2000	94	86	52	
2001	110	45	63	109
2002	140	43	62	112
2003	108	44	60	125
2004	133	64	83	106
2005	113	76	101	141
2006	139	67	128	152
2007	131	67	196	101
2008	134	110	136	144
TOTAL	1,498	783	1,004	990

using this index. We also determined the other set of representative terms using the temporal pattern extraction method based on the tf-idf values for the annual sets of documents.

Then, we perform a comparison of the document clustering based on the two sets of representative terms for each conference. The clustering result criterion is measured by the χ^2 values between the clusters and the published years. This indicates the availability to distinguish the trends related to the research interests with considering the time differences.

3.1 Extracting Technical Terms

Technical terms were determined in the sets of documents using the term extraction method [11][3]. Considering the difficulties of the term extraction without a dictionary, a term extraction method based on the adjacent frequency of compound nouns was applied. This method involves the detection of technical terms by using the following values for each candidate CN:

$$FLR(CN) = f(CN) \times (\prod_{i=1}^{L}(FL(N_i)+1)(FR(N_i)+1))^{\frac{1}{2L}}$$

where $f(CN)$ denotes the frequency of a candidate noun CN, and $FL(N_i)$ and $FR(N_i)$ denote the frequencies of different words to the right and the left of each noun N_i in bi-grams including each CN. In the experiments, technical terms were extracted using this FLR score as $FLR(t) > 1.0$. Simple stop word elimination is used to determine

[3] The implementation of this term extraction method is distributed in `http://gensen.dl.itc.u-tokyo.ac.jp/termextract.html` (in Japanese).

the termination of CN,. Note no stemming techniques were used because we want to consider detailed differences in the terms.

The extract technical terms were extracted using the term extraction method, shown in Table 2. After merging all of the titles of each conference into one set of documents, these terms were extracted for each set of titles.

Table 2. Description of the numbers of extracted terms

	KDD	PKDD	PAKDD	ICDM
# of extracted terms	1,941	1,011	1,400	1,948

3.2 Representative Terms Based on the tf-idf for All of the Documents

In order to select representative terms, for all of the documents D_{all}, we calculated the tf-idf values of terms $term_i$ as follows:

$$tfidf(term_i, D_{all}) = df(term_i, D_{all}) \times log \frac{|D_{all}|}{df(term_i, D_{all})}$$

where $tf(term_i, D_{all})$ is the frequency of each term $term_i$ in all of the sets of documents $|D_{all}|$. Similarly, $df(term_i, D_{all})$ is the frequency of documents d_x containing each term $term_i$. $|D_{all}|$ denotes the total number of documents for each conference.

Table 3 shows the top ten terms based on the tf-idf values.

Table 3. Top 10 selected terms based on the tf-idf index in the titles of the four conferences

KDD		PKDD		PAKDD		ICDM	
$term_i$	$tfidf(term_i)$	$term_i$	$tfidf(term_i)$	$term_i$	$tfidf(term_i)$	$term_i$	$tfidf(term_i)$
mining	495.8	data	224.8	mining	319.3	mining	331.9
data	488.7	mining	219.8	data	287.2	data	316.1
using	325.9	learning	161.8	using	249.2	clustering	249.9
data mining	274.4	using	161.8	based	236.8	using	238.1
learning	260.2	discovery	158.8	clustering	228.4	learning	197.7
discovery	248.6	clustering	137.6	algorithm	182.4	classification	185.4
knowledge	234.5	classification	134.0	efficient	170.9	algorithm	170.0
web	219.6	association	125.8	association	169.0	approach	162.7
clustering	208.2	knowledge	121.9	patterns	152.0	patterns	154.2
efficient	177.1	rules	112.9	frequent	151.0	text	148.3

For the representative terms for document clustering, the top 10% of the terms were selected based on the tf-idf value. Each document d_x is represented as the Boolean vector $\mathbf{x} = (bx_1, bx_2, ..., bx_p)$, where bx_i is represented as follows:

$$bx_i = \begin{cases} 1 & (term_i \in d_x) \\ 0 & (term_i \notin d_x) \end{cases} \tag{1}$$

We used 194 terms for KDD, 101 terms for PKDD, 140 terms for PAKDD, and 195 terms for ICDM as the feature terms in the following document clustering. We denote this set of features as "Feature Set I".

3.3 Extracting Temporal Patterns of Term Importance Index by Using K-Means Clustering

For each set of documents $(D_1, D_2, ..., D_m)$, the following tf-idf values were calculated for each period t_j:

$$tfidf(term_i, D_{t_j}) = tf(term_i, D_{t_j}) \times log\frac{|D_{t_j}|}{df(term_i, D_{t_j})}$$

where $tf(term_i)$ is the frequency of each term $term_i$ in the corpus with $|D_{t_j}|$ documents. $|D_{t_j}|$ is the number of documents included in each period. Using this definition, we can arbitrarily specify the length of the period. Annual articles ware used in this experiment, therefore a period of year is assigned for each set. $df(term_i, D_{t_j})$ is the frequency of documents containing $term_i$.

To the datasets consisting of temporal values of the importance indices, we extract temporal patterns by using k-means clustering implemented in Weka [16]. We set up the numbers of one percent of the terms as the maximum number of clusters k for each dataset. Then, the system obtained the clusters with minimizing the sum of squared error within clusters. By iterating less than 500 times, the system obtains the clusters by using Euclidian distance between two sequences $\mathbf{x}$ and $\mathbf{y}$ which is defined the following[4]:

$$Sim(\mathbf{x}, \mathbf{y}) = \sqrt{\sum_{j=1}^{m}(x_i - y_i)^2}$$

Table 4 shows the result of k-means clustering. The sum of squared error is calculated as follows:

$$SSE(D) = \sum_k \sum_i \sum_j (v_{it_j} - c_{kt_j})^2$$

where c_{kt_j} is the centroid value of c_k for each time-point t_j.

Table 4. Sum of squared errors of the technical terms clustering for the titles of the four conferences

KDD		PKDD		PAKDD		ICDM	
# of clusters	SSE	# of clusters	SSE	# of clusters	SSE	# of clusters	SSE
12	52.3	10	45.6	10	34.8	11	20.7

For the temporal patterns of term usages, we extracted the temporal clusters of the titles for each conference. Table 5, Table 6, Table 7, and Table 8 show the numbers of clusters c_k and the representative terms of each cluster c_k respectively. The representative term of each cluster c_k has the highest FLR score within the cluster.

For the representative terms for document clustering, we characterize each document d_x by using the temporal patterns c_k and all of the terms $term_i$. Each document d_x is

[4] The system also normalized the values for each year.

Table 5. The numbers of terms included in the temporal patterns and the representative terms of each temporal pattern on the titles of KDD from 1994 to 2008

k	# of $term_i \in c_k$	representative term c_k
1	114	high-dimensional data
2	145	data sets
3	195	relational data
4	137	web usage mining
5	99	distributed data mining
6	156	data streams
7	741	database mining
8	191	graph mining
9	91	learning bayesian networks
10	49	knowledge discovery
11	16	clustering
12	7	data mining

Table 6. The numbers of terms included in the temporal patterns and the representative terms of each temporal pattern on the titles of PKDD from 1997 to 2008

k	# of $term_i \in c_k$	representative term c_k
1	29	knowledge discovery
2	108	relational learning
3	12	data mining
4	92	spatial data
5	348	temporal data mining
6	98	data analysis
7	83	data clustering
8	57	text mining
9	92	data streams
10	92	application of data mining

represented as the Boolean vector $\mathbf{x} = (bx_1, bx_2, ..., bx_p)$, where bx_k is represented as the following:

$$bx_k = \begin{cases} 1 & (term_i \in c_k \cap term_i \in d_x) \\ 0 & (Otherwise) \end{cases} \tag{2}$$

Thus, we constructed the features of the document consisting of information on the temporal clusters. The number of features for KDD, PKDD, PAKDD, and ICDM is 12, 10, 10, and 11, respectively. We denote this set of features as "Feature Set II".

3.4 Comparing Document Clustering Results by Using the Two Feature Sets

The results of the document clustering for d_x of all of the titles for each conference D_{all} were compared using the two feature sets for each conference, as described in sections 3.2 and 3.3. In order to compare the dependency between the document clusters based

Table 7. The numbers of terms included in the temporal patterns and the representative terms of each temporal pattern on the titles of PAKDD from 1998 to 2008

k	# of $term_i \in c_k$	representative term c_k
1	208	frequent pattern mining
2	8	clustering mining
3	70	structured association patterns
4	179	hierarchical clustering based
5	83	data clustering
6	111	association rule mining
7	27	knowledge discovery
8	612	data mining based
9	97	text mining
10	14	data mining

Table 8. The numbers of terms included in the temporal patterns and the representative terms of each temporal pattern on the titles of ICDM from 2001 to 2008

k	# of $term_i \in c_k$	representative term c_k
1	206	knowledge discovery
2	43	text mining
3	653	data clustering
4	286	web mining
5	265	data mining approach
6	13	data mining
7	29	association rules
8	266	spatial data mining
9	5	mining
10	32	text classification
11	147	data analysis

on the Boolean vectors of d_x and the published years t_j, the following χ^2 statistics were compared:

$$\chi^2 = \sum_K \sum_{t_j} \frac{(F_{Kt_j} - f_{Kt_j})^2}{f_{Kt_j}}$$

where F_{Kt_j} is the observed number of the documents included in the document cluster C_K and the set of document in t_j as defined $|d_x \in C_K \cap d_x \in D_{t_j}|$, and f_{Kt_j} is the expected value.

The k-means clustering algorithm was used to extracted the document cluster C_K for all of the documents D_{all} for the titles of each conferences. K is set to 0.5% of the numbers of the automatically extracted terms. The K for KDD is 10, 5 for PKDD, 7 for PAKDD, and 10 for ICDM.

Table 9 shows the χ^2 statistics for the document clustering results. The χ^2 statistics for the document clusters achieved by Feature Set II, which consists of the appearances of terms included in the temporal patterns c_k, are more than 10 times higher that achieved by Feature Set I. Although the features of Feature Set I, which are the appearances of the representative terms selected by tf-idf values for all of the documents, are used 10 times more numbers than Feature Set II, Feature Set II yielded the document clusters with greater time dependency compared to the amount of the terms.

Table 9. χ^2 statistics between published years and the document clusters for KDD, PKDD, PAKDD, and ICDM titles

	Feature Set I (from 3.2)	Feature Set II (from 3.3)
KDD	193.4	1534.3
PKDD	54.7	723.1
PAKDD	83.4	894.5
ICDM	56.6	1228.9

Fig. 6 illustrates the document clustering results using the two feature sets of the paper titles of the four conferences. As shown in Fig. 6(a), Fig. 6(c), Fig. 6(e), and Fig. 6(g) the document clusters of the representative term appearances have small differences between each annual set of documents. However, the document clusters of the appearances of terms included the temporal patterns show differences between each annual set of documents as shown in Fig. 6(b), Fig. 6(d), Fig. 6(f), and Fig. 6(h).

This result may seem slightly trivial, considering that the features of Feature Set II are obtained by temporal patterns. However, this result cannot be obtained if the temporal patterns have no timeline features. Thus, the feature set consisting of the groups of terms based on the temporal behavior of the importance index indicates that the features can characterize the temporal differences of the published documents. The obtained document clusters then contain some similar temporal context based on the similar terms included in each temporal pattern.

4 Conclusion

In this chapter, we described the feature construction method, considering the temporal context based on the temporal patterns of term usages for categorizing temporally published documents. We implemented the method using the tf-idf index and k-means clustering for extracting temporal patterns as term usages for characterizing each document as a Boolean vector representation. The method was compared to document clustering using top-k feature terms selected by ranking the tf-idf values of the overall terms and the Boolean vector space model. Then, the dependencies between their published years and the document clustering result were measured by using χ^2 statistics.

The empirical results show that the time dependency of the document clusters achieved greater than 10 times higher χ^2 statistics for the four set of titles from the four datamining related conferences using our proposed method. The time dependency of the document cluster for ICDM titles improved from $p = 0.6$ using the conventional

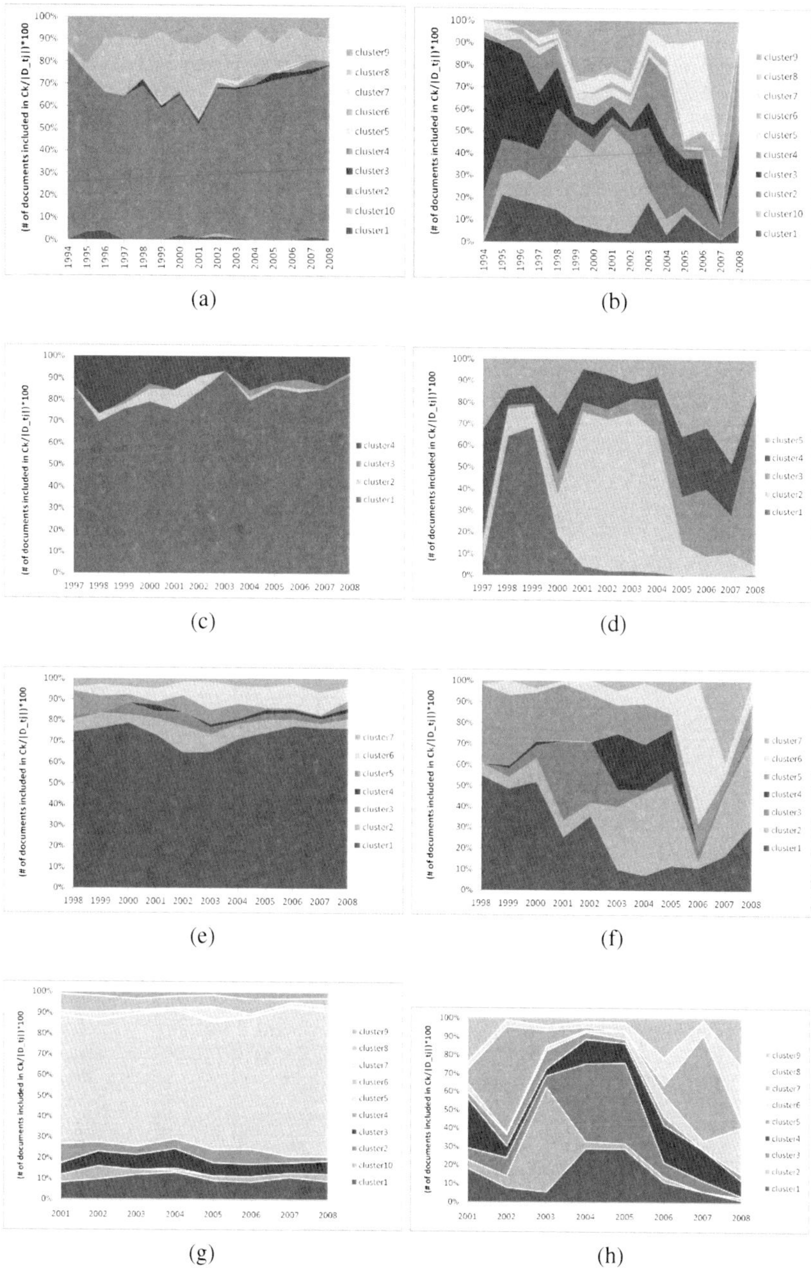

Fig. 6. The percentages of the assigned document clustering clusters based on the appearances of representative terms in (a)(c)(e)(g) and the appearances of terms included in the temporal patterns c_k in (b)(d)(f)(h). The bottom is 'cluster1', and the top is 'cluster9' sorted by dictionary order.

top-k bag-of-words approach to $p < 0.001$ using our method. At the same time, the groups having the similar temporal behaviors based on the importance index also give valuable information about temporal context as the similar terms in each group.

In the future, we will introduce temporal pattern recognition methods [12] and other temporal pattern extraction methods[5,8] to better represent for temporal information. We will also compare the efficiency of the combinations of indices and temporal pattern extraction methods for finding remarkable characteristics of temporally generated sequential data not only documents but event sequence in various situation such as finance, politics, security issue, and so forth.

References

1. The dblp computer science bibliography, `http://www.informatik.uni-trier.de/~ley/db/`
2. Abe, H., Tsumoto, S.: Text categorization with considering temporal patterns of term usages. In: Fan, W., Hsu, W., Webb, G.I., Liu, B., Zhang, C., Gunopulos, D., Wu, X. (eds.) ICDM Workshops, pp. 800–807. IEEE Computer Society (2010)
3. Anderberg, M.R.: Cluster Analysis for Applications. Monographs and Textbooks on Probability and Mathematical Statistics. Academic Press, Inc., New York (1973)
4. Dhillon, I.S.: Co-clustering documents and words using bipartite spectral graph partitioning. In: KDD 2001: Proceedings of the Seventh ACM SIGKDD International Conference on Knowledge Discovery and Data Mining, pp. 269–274. ACM, New York (2001)
5. Keogh, E., Chu, S., Hart, D., Pazzani, M.: Segmenting time series: A survey and novel approach. In: Data mining in Time Series Databases, pp. 1–22. World Scientific (2003) (an Edited Volume)
6. Kontostathis, A., Galitsky, L., Pottenger, W.M., Roy, S., Phelps, D.J.: A survey of emerging trend detection in textual data mining. A Comprehensive Survey of Text Mining (2003)
7. Lent, B., Agrawal, R., Srikant, R.: Discovering trends in text databases. In: KDD 1997: Proceedings of the third ACM SIGKDD International Conference on Knowledge Discovery in Data Mining, pp. 227–230. AAAI Press (1997)
8. Liao, T.W.: Clustering of time series data: a survey. Pattern Recognition 38, 1857–1874 (2005)
9. Lin, D., Wu, X.: Phrase clustering for discriminative learning. In: ACL-IJCNLP 2009: Proceedings of the Joint Conference of the 47th Annual Meeting of the ACL and the 4th International Joint Conference on Natural Language Processing of the AFNLP, vol. 2, pp. 1030–1038. Association for Computational Linguistics, USA (2009)
10. Mei, Q., Zhai, C.: Discovering evolutionary theme patterns from text: an exploration of temporal text mining. In: KDD 2005: Proceedings of the Eleventh ACM SIGKDD International Conference on Knowledge Discovery in Data Mining, pp. 198–207. ACM, USA (2005)
11. Nakagawa, H.: Automatic term recognition based on statistics of compound nouns. Terminology 6(2), 195–210 (2000)
12. Ohsaki, M., Abe, H., Yamaguchi, T.: Numerical time-series pattern extraction based on irregular piecewise aggregate approximation and gradient specification. New Generation Comput. 25(3), 213–222 (2007)
13. Ohsawa, Y., McBurney, P.: Chance discovery. Advanced information processing. Springer (2003)
14. Shaparenko, B., Caruana, R., Gehrke, J., Joachims, T.: Identifying temporal patterns and key players in document collections. In: IEEE ICDM Workshop on Temporal Data Mining: Algorithms, Theory and Applications (TDM 2005), pp. 165–174 (2005)

15. Sparck Jones K.: A statistical interpretation of term specificity and its application in retrieval. Document Retrieval Systems, 132–142 (1988)
16. Witten, I.H., Frank, E.: Data Mining: Practical Machine Learning Tools and Techniques with Java Implementations. Morgan Kaufmann (2000)
17. Yang, Y., Pedersen, J.O.: A comparative study on feature selection in text categorization. In: ICML 1997: Proceedings of the Fourteenth International Conference on Machine Learning, pp. 412–420. Morgan Kaufmann Publishers Inc., San Francisco (1997)
18. Zhao, Y., Karypis, G.: Empirical and theoretical comparisons of selected criterion functions for document clustering. Mach. Learn. 55(3), 311–331 (2004)

Finding Rare Patterns with Weak Correlation Constraint: Progress in Indicative and Chance Patterns

Yoshiaki Okubo, Makoto Haraguchi, and Takeshi Nakajima

Graduate School of Information Science and Technology
Hokkaido University
N-14 W-9, Sapporo 060-0814, Japan
{mh,yoshiaki}@ist.hokudai.ac.jp

Abstract. A notion of *rare patterns* has been recently paid attention in several research fields including *Chance Discovery*, *Formal Concept Analysis* and *Data Mining*. In this paper, we overview the progress of our investigations on rare patterns satisfying a *weak-correlation* constraint. A rare pattern must indicate some *significance* as well as a fact that the number of its instances is a few. We pay our attention to a pattern as an itemset in a transaction database which consists of several general items, but has a very small degree of correlation in spite of the generality of component items. Such a pattern is called an *indicative pattern* and is regarded as a rare pattern to be extracted.

In order to exclude trivial patterns of general items with few instances, we introduce an objective function for taking into account both the generality of component items and the number of instances as objective evidences. Then we try to find indicative patterns with the Top-N evaluation values under a constraint that the degree of correlation must not exceed a given upper bound.

For making a hidden relationship between a pair of more frequent patterns visible, the framework of finding Top-N indicative patterns is then extended by imposing some structural constraints to our indicative pattern and larger patterns bridged by it. As a recent progress in this direction, we briefly present a framework of finding *chance patterns* with *KeyGraph*®-based importance as well as some experimental result.

1 Introduction

In this paper, we overview the progress of our investigations on *rare patterns* satisfying a *weak-correlation* constraint which have been originated from [25].

1.1 From Frequent Patterns to Rare Patterns

In these decades, much attention has been paid for the study of finding major clusters in *clustering* [5] or frequent patterns in *pattern mining* [1,2,4]. Both the majority and the frequentness mean that the number of instances belonging to a

Y. Ohsawa & A. Abe (Eds.): Advances in Chance Discovery, SCI 423, pp. 231–246.
springerlink.com

cluster or including a pattern is many, at least not a few. Although those studies for major and frequent patterns are needless to say proven to be effective and useful, infrequent patterns or clusters with less number of instances are attracting our attention, as we see in the previous studies on *Chance Discovery* [9,11] and *rare patterns/concepts* [6,7,8,25,33].

A minor pattern is not necessarily a rare pattern. A rare pattern must indicate some significance in addition to a fact that the number of instances is a few. In case of [11], a candidate *dark event*, E, corresponds to a rare pattern in the following sense:

Minority: E has only a few instances.
Correlation with Major Cluster: It is correlated to some major clusters, although the degree of correlation is small.
Potential for Chance: E has some potential for becoming a chance, infrequent but visible event correlated with major clusters of events.

The latter two above introduce a kind of significance of minor event.

In case of [25,33], on the other hand, a rare pattern is defined as a *compound event* which is a set of items as primitive events, in a transaction database. More concretely speaking, a rare pattern P is required to be *concise* in the sense that P consists of *less number of more general items.* By the conciseness, it is easier to read and understand the output rare patterns. In addition to this fact, it should be noted here that general items have higher probability for being connected to (or co-occured with) another ones. In spite of the potential connectivity, the number of instances becomes fewer after combining them as a compound event. We consider this kind of phenomenon opposite to a notion of *modularity* [14,15], and regard such patterns as significant ones. As a special case, a dark event E may be realized by combining general events (items) $e_1, ..., e_k$, provided there exists a highly correlated itemset C whose instance set intersects with one for E a little. As each e_i is general, E has potential for obtaining more intersection with C as instance sets in some future database.

1.2 Indicative Concepts

The notion of concise (rare) pattern consisting of a small number of general items has been revised in [30] by taking *weak-correlatedness* and *objective evidence* of patterns into account.

Weak-Correlatedness of Patterns: A pattern P must have lower correlation, where as P may consist of more than three items, we use *Bond* measure [16], an extension of *Jaccard Coefficient*, in order to evaluate correlation.

The conciseness introduced in [33] does not cover the case of combination of correlated patterns. So we have improved it so that our target pattern may involve a pair of correlated patterns whose overall correlation after combining them dramatically decreased. Moreover, since a pattern with lower correlation has in general only a small number of instances, the constraint on weak-correlatedness can be understood as a weaker version of minority constraint.

Objective Evidence under Correlation Constraint: As we prefer a combination of more general items as long as the combination results in a fewer instances, it is often the case that the number of instances and Bond measure are zero. So, we consider that a rare pattern with more evidences is superior to another ones with less evidences.

As a result, we prefer a concise pattern P of more general items and with more instances, as long as P satisfies the correlation constraint.

To make our arguments simple, we introduce an integrated monotonic objective function that is a linear combination of generality of component items and the number of evidences (instances). The function is used to evaluate tentative patters w. r. t. the generality of component items and the number of instances. The optimization task is to maximize the function value in a pattern lattice subject to the correlation constraint. For the sake of computational efficiency, we consider so called *closed patterns* [24] as representatives of patterns with the same supports, and evaluate their correlation after removing redundant items. A closed pattern P is called an *indicative pattern* if it has more evidences and more generality of component items among patterns satisfying the correlation constraint, where as a closed pattern is an equivalent notion of the *intent* of a *formal concept* [23], P is also called an *indicative concept.*

There exist a lot of locally optimal patterns satisfying the constraints. By this reason, instead of applying a greedy method, we propose to extend a *Top-N Branch-and-Bound Search Method* for finding minor closed patterns [26,27], where N is a given threshold designating how many number of solutions must be presented to users. It tries to combine instance objects to form closed patterns, and it always keeps tentative top-N values of objective function. It stops adding more instances to the present closed pattern, whenever it turns out that the estimated evaluation value of the present pattern after adding every possible instances cannot exceed the least value of tentative top-N values. This is a kind of *safe* branch-and-bound search control because of the monotonicity of Bond measure and the objective function as well.

In our experimentation, we can verify that our algorithm can extract an interesting indicative concept from a collection of newspaper articles. For example, we can obtain a pattern appeared in just 5 articles in 1994-collection, but it can be found in 115 articles in 1995 collection. In addition to the frequency change, the contents are also drastically changed due to "The South Hyogo Prefecture Earthquake in 1995". Before the earthquake, the pattern is appeared in articles with various topics including, e. g., a designer competition and a plan for business expansion. After the earthquake, however, most of the articles in which the pattern appeared are concerned with information of medical, donations for relief and daily life.

In *Contrast Set Mining* [17], a family of transaction databases are given to find out a pattern P distinguishing those databases. For at least one pair of databases DB_1 and DB_2, P is required to be more minor in DB_1 and more frequent in DB_2. In addition, P must be never statistically independent with DB, where χ^2 test is applied to judge if the null hypothesis of independence can be rejected. However, as is well-known, χ^2 statistics is not normally applied to patterns with very fewer instances, and is therefore not adequate for our purpose of finding indicative patterns.

Similar to the notion of contrast sets, a pattern whose support is low in one database and higher in another database is regarded as a *Emerging Pattern* [18,19,20]. However our concern is the correlation, not a support notion.

Another similar notion of emergingness can be found in [28]. A correlation change is defined as a ratio of *Lifts* [21] over two databases. Then the task is to find patterns with the higher ratio of Lifts. Particularly the study aims at finding a pattern with very low Lift in one database and relatively higher Lift in another database. However, as the study of Brin [22] has suggested, the Lift measure is in some cases misleading from the viewpoint of correlation.

It would be also possible to consider the emergingness w.r.t. correlation in the sense of Bond or Jaccard Coefficient. Although we leave a design of algorithm customized for the correlation change as a future work, as has been mentioned just above, we have observed in [30] that the correlation of some meaningful indicative pattern in one database is actually increased in another database. This suggests a possibility of the next stage algorithm for realizing correlation change detectors. As a closely-related work, a framework for detecting change of *information-based correlation* over two databases has been discussed in [29].

1.3 Indicative Patterns Connecting Larger Patterns

Although we have found some interesting indicative patterns in [30], we need another criterion to be imposed on potential patterns. This is simply because a pattern with very low frequency (support) may happen *accidentally*. About this issue, the study of *Chance Discovery* [9,11] should be referred to again here. Chance discovery also aims at finding rare patterns, where the underlying representation is a unigraph of feature attributes linked with correlation weights. Then a chance is an attribute node that has some correlation to clusters of attributes. When we consider a rare pattern of attributes and register it as a new node in the graph, it can be regarded as a dark event in [11]. In order to find such dark events, some heuristic method using *Simulated Annealing* has been tried in [11], where the finding process involves complicated re-clustering operations and many user-interactions to select clusters from user's viewpoint. In [32], therefore, we have tried to formulate the problem of extracting dark events in terms of *concept*

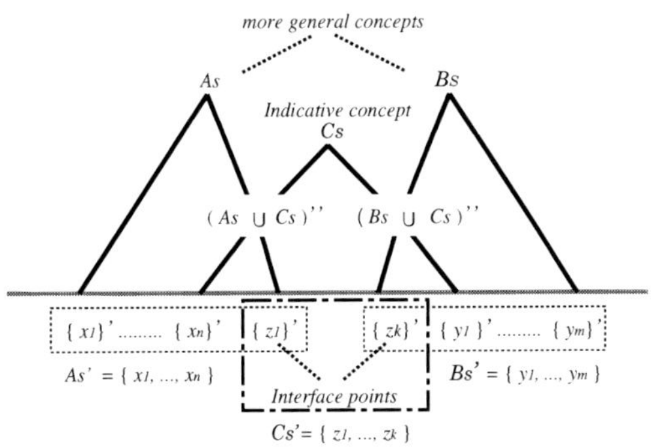

Fig. 1. Structural Constraint for Indicative Pattern

lattice [23] so that we can present a systematic search procedure for finding indicative closed pattern under a *structural constraint* depicted by the Fig. 1.

In the figure, As, Bs, Cs are intents (that is, closed patterns), the prime symbol "$'$" denotes a derivation operator that defines *Galois connection*, and x_i, y_j, z_k are objects (instances). Under this notation, our target indicative pattern Cs is required to be a common superconcept of some subconcepts of concepts As and Bs with larger extents (instances). As As and Bs are more general, they can have various aspects constrained by attributes in Cs. As Cs have larger extent, those aspects will be visible but are less interesting as everyone can easily recognize them. As Cs are more indicative and rare, it will make us realize some *hidden structural connection* between As and Bs.

Unlike a standard procedure for enumerating (closed) patterns, our algorithm firstly enumerates indicative patterns Cs, while checking the existence of more frequent As and Bs, given Cs. In a few words, every tentative pattern is evaluated by an objective function regarding the intensional generality of concepts and their extent sizes, where the former factor is involved in the definition of indicative pattern and the latter for excluding too small (minor) concepts. Then, our enumerator tries to find only top-N patterns w.r.t. the objective function under the constraint that the degree of correlation must not exceed a given threshold in order to keep weak-correlatedness of patterns. For each tentative top N indicative pattern Cs, we check the existence of As and Bs. The interface points in Fig. 1 is selectively chosen from the extent of Cs as staring objects to form As and Bs, respectively. In addition to this combination, there exist many ways to have possible extents for As and Bs. We, therefore, further introduce some constraints to exclude meaningless As and Bs. The first criterion is concerned with distance between As and Bs.

Farness of More Frequent Concepts to be Connected: We require that As and Bs are "far away". If so, the degree of unexpectedness for them to be connected by the indicative pattern will be increased. Actually we set a parameter about distance and require that the distance must be greater than it. Consequently they are at least extensionally disjoint.

Although many frequent concepts are pruned away by the structural constraint and the farness constraints, there may still remain many concepts that can be candidates for As and Bs. To further reduce the possibility and to make them more understandable, we introduce a measure for *conceptual clarity* and present to users only top N candidates according to it. The final selection is thus made by users based on their interests. Unlike the original work on dark events, user interaction is thus required as a post-processing.

Conceptual Clarity of Concepts to be Connected: The same objective function is also used to search for As and Bs so as to obtain patterns of more general attributes for understandability. In addition, we require that each of As and Bs consists of only objects with similar attributes in order to retain conceptual clarity.

Note that the similarity is just the degree of overlappingness of attribute sets those objects have. So, we simply use the Bond measure for objects, instead of Bond for attributes to calculate the correlation among attributes.

According to the framework, for a collection of newspaper articles in 1994 with a keyword "Kobe" (a name of city in Japan), we have found an indicative concept (pattern) connecting "immigration control" and "Soccer". The indicative pattern describes a family of newspaper articles about "various sports games held in Tokyo by representative players". Although "immigration control" and "Soccer" seems conceptually different from each other, they can be connected by the indicative pattern. For example, the concept "Soccer" includes an article about national soccer team players of some country who flied to Japan to attend a game in Tokyo. It is, therefore, concerned with immigration control process. Thus, our indicative concept has an ability to detect such an interesting relationship between concepts which seem to be conceptually far apart at first glance. We can expect that our indicative concepts would enhance our chance to detect unexpected surprising information or knowledge.

1.4 Towards Chance Patterns with *KeyGraph*®-Based Importance

In the framework in [32], we have to basically find *triples* of patterns, an indicative pattern and a pair of connected ones, in a concept space which is in general quite huge. In order to define a problem of finding *meaningful* triples of concepts which can be solved *exactly* (not *heuristically*), therefore, we are required to impose several constraints. As has been discussed just above, the problem formulation in [32] seems a little bit complex. In [31], therefore, we have reformulated our problem of finding an indicative pattern which connects more frequent patterns by taking a *KeyGraph*®-based importance of patterns.

KeyGraph® [1] has been originally proposed in the literature [10] as a useful algorithm for indexing a document and is widely-accepted in the field of Chance Discovery. The underlying idea has been applied to many actual domains (e.g. [11,12]). In a word, given a document to be indexed, *KeyGraph*® identifies a set of key words each of which co-occurs with some frequent terms. It should be noted here that each of the extracted key words is *not necessarily* frequent. They are rather rare but significant in the document. This kind of importance is considered in our problem of finding indicative patterns connecting more frequent patterns, regarding a document and a word as a transaction and an item, respectively. In previous *KeyGraph*®-based frameworks, each target to be extracted is supposed to be an individual word or event. In our framework, on the other hand, our target could be a combination of words or a compound event.

Our pattern (of words) to be extracted is defined as an indicative pattern C which is *supported* by a pair of frequent patterns X_L and X_R in the following sense: *"The pattern C often appears in a part of documents containing X_L, and also in those containing X_R"*. More technically speaking, this means that the

[1] *KeyGraph*® is a registered trademark owned by Yukio OHSAWA, The University of Tokyo, JAPAN.

confidence values of *association rules*, $X_L \Rightarrow C$ and $X_R \Rightarrow C$, are relatively high. In this sense, some relationship between X_L and X_R is *implied* by C. Moreover, by imposing a constraint for *weak-correlatedness* of words in C based on the Bond measure, we can also expect C to be rare. The Bond measure is also used for evaluating (conceptual) similarity among documents containing C, as is like in [32]. We call such an indicative pattern C a *chance pattern*. It should be emphasized here that as opposed to the framework in [32], we do not require that C must be a concept in our concept lattice. That is, our chance pattern could be just a pattern, not a closed pattern. In this sense, our chance pattern might be invisible in the concept lattice. Therefore, our chance of finding hidden and implicit relationships between X_L and X_R would be further enhanced.

The rest of this paper is organized as follows. The next section introduces some basic terminologies. Our notion of indicative patterns/concepts is presented in Section 3 and then extended to the notion of chance patterns in Section 4. We briefly present our computation procedure for extracting chance patterns as well as some experimental result. This paper is concluded in Section 5 with a summary and future works.

2 Preliminaries

Formal Concept Analysis is well known as a useful framework for conceptual data analysis [23]. Let $\mathcal{O}$ be a set of *objects* (or individuals) and $\mathcal{A}$ a set of *attributes* (or features). For a binary relation $R \subseteq \mathcal{O} \times \mathcal{A}$, a triple $(\mathcal{O}, \mathcal{A}, R)$ is called a *formal context*. If $(x, y) \in R$, we say that the object x has the attribute y or y is associated with x. Then, for an object $x \in O$, the set of attributes associated with x is denoted by $Att_R(x)$, that is, $Att_R(x) = \{y \in \mathcal{A} \mid (x, y) \in R\}$.

Given a formal context $(\mathcal{O}, \mathcal{A}, R)$, for a set of objects $X \subseteq \mathcal{O}$ and a set of attributes $Y \subseteq \mathcal{A}$, we define two mappings $\varphi : 2^{\mathcal{O}} \to 2^{\mathcal{A}}$ and $\psi : 2^{\mathcal{A}} \to 2^{\mathcal{O}}$ as follows:

$$\varphi(X) = \{y \in \mathcal{A} \mid \forall x \in X, y \in Att_R(x)\} = \bigcap_{x \in X} Att_R(x).$$

$$\psi(Y) = \{x \in \mathcal{O} \mid Y \subseteq Att_R(x)\}.$$

The former computes the set of attributes shared by every object in X. The latter, on the other hand, returns the set of objects with Y.

Based on these mappings, for a set of objects $X \subseteq \mathcal{O}$ and a set of attributes $Y \subseteq \mathcal{A}$, the pair of X and Y, (X, Y), is called a *formal concept* under the formal context if and only if $\varphi(X) = Y$ and $\psi(Y) = X$, where X and Y are called the *extent* and the *intent* of the concept, respectively. From the definition, it is easy to see that $\psi(\varphi(X)) = X$ and $\varphi(\psi(Y)) = Y$. That is, a formal concept is defined as a pair of *closed* sets of objects and attributes under the mappings.

For a formal context $(\mathcal{O}, \mathcal{A}, R)$, the binary relation R can be regarded as a transaction database in *Data Mining*, where each object and attribute correspond to a transaction and an item, respectively. Then, for a formal concept

(X, Y), the intent Y is equivalent to a *closed itemset* [24] and the extent X the set of transactions containing the itemset. That is, the size of X, $|X|$, gives the *frequency* of Y, denoted by $freq(Y)$, in the corresponding database.

3 Indicative Concepts with Weakly-Correlated General Intents

In this section, we present a notion of *Indicative Concepts* [30] which is an improved version of *Concise Rare Concepts* discussed in [25,33].

As has been previously discussed, we pay our attention to a concept with the intent consisting of *weakly-correlated general attributes.* By the weak-correlatedness, we expect such a concept tends to be *rare* because the combination of those attributes would be infrequently observed in general. By the generality, furthermore, it is expected that the concept has a clear meaning understandable for us.

3.1 Weak-Correlatedness of Intent

Given a concept, we evaluate the correlatedness of the intent (i.e. a pattern) by *Bond* [16].

Definition 1. (Bond-Based Correlatedness of Intent)
For a concept (X, Y), the correlatedness of the intent, denoted by $correl(Y)$, is defined as

$$correl(Y) = \frac{\mid \bigcap_{a \in Y} \psi(\{a\}) \mid}{\mid \bigcup_{a \in Y} \psi(\{a\}) \mid}. \qquad \blacksquare$$

In general, for a concept (X, Y), there exists Z such that $Z \subseteq Y$ and $\psi(Y) = \psi(Z) = X$, where we call Z a *generator* of Y. In other words, each attribute in $Y \backslash Z$ is *redundant* for identifying X. From the definition, since $correl(Z) \geq correl(Y)$ holds, we are required to remove those redundant features in order to evaluate its weakly-correlatedness more exactly. However, since there exist several generators of Y which are minimal under set-inclusion relation, computing all of the minimal ones is costly. Therefore, we consider here the following generator of Y as a *representative.*

Definition 2. (Representative Generator)
Let us assume $Y = \{a_1, \ldots, a_k\}$ and a_i is arranged in frequency-descending order. Then a *representative generator* of Y, denoted by $rep\_gen(Y)$, is defined as

$$rep\_gen(Y) = \{a_i \in Y | \psi(\{a_i\}) \not\supseteq \psi(\{a_{i+1}, \ldots, a_k\})\}. \qquad \blacksquare$$

$rep\_gen(Y)$ is the set of attributes each of which can never be implied by any combination of less frequent attributes. Intuitively speaking, highly frequent attributes tend to make bond values lower. It is better to remove such undesirable attributes if they are implied by some other attributes. The representative generator defined above is based on this idea. Then, we evaluate the correlatedness of the intent Y as $correl(rep\_gen(Y))$.

In order to define *weak-correlatedness* of intent, we impose a restriction on correlatedness by an upper bound σ. For a concept (X, Y), if $correl(rep\_gen(Y)) \leq \sigma$, then the intent is considered weakly-correlated and it becomes a candidate of indicative concepts.

It should be noted here that as has been reported in [30], even if our attention is paid only to such restricted concepts based on weak-correlatedness, the number of them is still quite huge. For example, for a dataset (formal context) consisting of 2,343-Japanese news articles (as objects) with 30,085-nouns (as attributes), we often have a large number of concepts with the order of million under various σ thresholds. It would be, therefore, impractical and intractable to extract all of concepts satisfying the weak-correlatedness constraint. We are highly required to reduce the number of concepts to be detected.

3.2 Generality of Intent

Since concepts are interpreted based on their intents, it would be difficult to provide an adequate meaning of a concept whose intent consists of *specific* attributes because those attributes make the meaning of the concept unclear. In order to obtain easily understandable concepts, we take *generality* of intents into account.

Definition 3. (Generality of Intent)
Let (X, Y) be a concept. Then, the *generality* of the intent Y, denoted by $generality(Y)$, is defined as

$$generality(Y) = \min_{a \in Y}\{|\psi(\{a\})|\} = \min_{a \in Y}\{freq(\{a\})\}. \qquad \blacksquare$$

If the intent consists of only general attributes, its generality becomes high. It is expected that we can clearly interpret concepts with such intents.

3.3 Objective Evidence of Intent

In general, weakly-correlated intents would have their corresponding extents with smaller size. In some case, a concept might have the extent consisting of few objects. However, such a concept seems to be undesirable, since its objective evidence is too weak and it might be regarded as just an *exception*. Therefore, we also take the size of extent into account as an *objective evidence* of intent.

Definition 4. (Objective Evidence of Intent)
For a concept (X, Y), the *objective evidence* of the intent Y, denoted by $evidence(Y)$, is defined as

$$evidence(Y) = |X| = freq(Y). \qquad \blacksquare$$

3.4 Top-N Indicative Concept Problem

We prefer a concept whose intent is weakly-correlated and general and has sufficient objective evidence. We call such a concept an *Indicative Concept.*

Due to a enormous number of formal concepts, we can easily imagine that the number of indicative concepts will be also large. Top-N approach [26,27], therefore, would be reasonable and practical.

Definition 5. (Top-N Indicative Concept Mining)
Let $(\mathcal{O}, \mathcal{A}, R)$ be a formal context, σ a weak-correlatedness threshold, α a weight of generality and β a weight of objective evidence.

Top-N Indicative Concept Mining is to extract formal concepts $C = (X, Y)$ such that

Weak-Correlatedness as Constraint:
$correl(rep\_gen(Y)) \leq \sigma$, and
Generality and Objective Evidence as Objective Function:
$eval(Y) = \alpha \cdot generality(Y) + \beta \cdot evidence(Y)$ is in the top-N among the concepts satisfying the weak-correlatedness constraint. ∎

More detailed discussion on indicative concepts, including a depth-first branch-and-bound algorithm for the problem and some experimental results, can be found in [30].

4 Towards Chance Patterns with *K*eyGraph®-Based Importance

The notion of indicative concepts has been extended in [32] so that we can make a hidden relationship between a pair of larger (more frequent) concepts visible. Then the problem has been recently reformulated into a problem of finding *Chance Patterns* with *KeyGraph*®-based importance [31]. In this section, we briefly present our current framework for finding chance patterns as well as a preliminary experimental result. In what follows, it is supposed that we are given a collection of documents, $\mathcal{D}$, each of which is represented as a set of words in a vocabulary $\mathcal{W}$. That is, we consider a pattern as a set of words.

Roughly speaking, a chance pattern is defined as an indicative pattern which can be *supported* by a pair of more frequent patterns, where for patterns X and Y, "X *supports* Y" means that Y often appears in the documents containing X. Such a chance pattern C supported by X_L and X_R would reveal some connection between X_L and X_R because the pattern C is shared with both a part of documents with X_L and a part of those with X_R. Particularly, the connection would be more interesting if X_L and X_R are conceptually different patterns.

4.1 Chance Pattern as Weakly-Correlated Words

Based on the idea in [30], we require a chance pattern must consist of *weakly-correlated general words.*

Let w_i and w_j be a pair of words in $\mathcal{W}$. As a special case of Def. 1, a *correlation between w_i and w_j*, denoted by $correl(w_i, w_j)$, is simply define as

$$correl(w_i, w_j) = correl(\{w_i, w_j\}) = \frac{|\psi(\{w_i\}) \cap \psi(\{w_j\})|}{|\psi(\{w_i\}) \cup \psi(\{w_j\})|}.$$

For any pair of words w_i and w_j in a pattern X, if $correl(w_i, w_j)$ is lower than a given threshold, then we consider the pattern X to be *weakly-correlated.*

As will be mentioned later, we also take *generality* of words into account. From the generality of the words, it is expected that such a pattern can be concisely interpreted.

4.2 Conceptual Similarity of Documents

Given a pattern X, the set of documents containing X, $\psi(X)$, is uniquely determined. In order to make the pattern meaningful, it would be reasonable to require that $\psi(X)$ consists of documents which are *conceptually similar.* Therefore, any pattern to be extracted is imposed a constraint for this requirement.

Let d_i and d_j be a pair of documents in $\mathcal{D}$. A *similarity between d_i and d_j* is defined as

$$sim(d_i, d_j) = \frac{|\varphi(\{d_i\}) \cap \varphi(\{d_j\})|}{|\varphi(\{d_i\}) \cup \varphi(\{d_j\})|}.$$

The numerator is the number of words appeared in both d_i and d_j and the dominator the number of those included in d_i or d_j (or both). Thus, a higher value of $sim(d_i, d_j)$ shows that most of the words in d_i or d_j are shared in both documents. That is, they are expected to be *conceptually similar.* Thus, the idea of *Bond* in [16] is used for both words and documents in our framework.

Based on the similarity, we can easily define a similarity among more than two documents. For a set of documents D, a *similarity among the documents in D*, denoted by $Sim(D)$, is defined as the minimum of the similarity between any pair of documents in D. That is,

$$Sim(D) = \min_{d_i, d_j \in D} \{sim(d_i, d_j)\}.$$

If D gives a higher value of Sim, we consider that D consists of conceptually similar documents.

4.3 Supporting Power of Base Patterns

Let X be a pattern of words in $\mathcal{W}$. Given a pattern $W \subseteq \mathcal{W}$, a *power of X for supporting W* is denoted by $power(X, W)$ and defined as

$$power(X, W) = \frac{|\psi(W) \cap \psi(X)|}{|\psi(X)|},$$

where X is called a *base pattern supporting W*. Roughly speaking, $power(X, W)$ corresponds to the conditional probability of W given X and is also known as the *confidence* of an association rule $X \Rightarrow W$ [3].

Let X_i and X_j be a pair of patterns in $\mathcal{W}$. Given a set of words $W \subseteq \mathcal{W}$, we simply define a *supporting power of* X_i *and* X_j *for* W, denoted by $SP(X_i, X_j, W)$, as the minimum of *power* of each base pattern, that is,

$$SP(X_i, X_j, W) = \min\{power(X_i, W), power(X_j, W)\}.$$

4.4 Problem of Mining Top-N Chance Patterns

Based on the above discussion, we can now define our problem of mining chance patterns each of which can be strongly supported by a pair of more frequent base patterns with a Top-N value of supporting power.

Definition 6. (Top-N Chance Pattern Mining)
Let $\mathcal{D}$ be a set of documents, max_correl an upper bound of $correl$ for weak-correlatedness of pattern and min_sim a lower bound of Sim for conceptual similarity within pattern. Then a problem of *mining Top-N chance patterns* is to extract triples of patterns (C, B_L, B_R) such that
(C1) Conceptual Similarity within Pattern:
$Sim(\psi(C)) \geq min\_sim$, $Sim(\psi(B_L)) \geq min\_sim$ and $Sim(\psi(B_R)) \geq min\_sim$.
(C2) Closedness of Base Patterns:
B_L and B_R are closed, that is, $\varphi(\psi(B_L)) = B_L$ and $\varphi(\psi(B_R)) = B_R$.
(C3) Minimality of Base Patterns:
B_L and B_R are minimal among patterns satisfying C1 and C2.
(C4) Disjointness of Base Patterns:
$\psi(B_L) \cap \psi(B_R) = \emptyset$.
(C5) Weak-Correlatedness within Chance Pattern:
$\max_{w_i, w_j \in C}\{correl(w_i, w_j)\} \leq max\_correl$.
Supporting Power of Base Patterns as Objective Function:
The value of $SP(B_L, B_R, C)$ is in the Top-N among triples of patterns satisfying the above five constraints, where C is called a *chance pattern*. ■

We impose five constraints on our patterns to be extracted. By the constraint C1, all of the documents with our pattern are expected to be conceptually similar. In this sense, our patterns are meaningful. The constraint C2 requires that our base pattern is the *intent* of a concept. Thus, our chance pattern intensionally connects conceptual groups of documents. The constraint C3 means that our base patterns must be as frequent as possible so that our chance pattern suggests a hidden relationship between general concepts we can easily observe. Interestingness of such a relationship would be further gained with the constraint C4 which requires the connected concepts to be extensionally disjoint, that is, non-overlapping. By the constraint C5, our chance patterns tend to be rare because the number of documents containing a combination of weakly-correlated words is generally expected to be small. Moreover, it should be noted that maximizing our objective function implicitly means we prefer a chance pattern which consists of general (frequent) words. It is, therefore, expected that our chance pattern can be interpreted concisely and clearly.

4.5 Algorithm Sketch

Target triples (C, B_L, B_R) are extracted according to the following three phases:

Enumerating Candidates of Base Patterns: We first enumerate the patterns which satisfy the constraints C1, C2 and C3. The set of such patterns is denoted by $Bases$ and used as candidates of base patterns. Actually, each candidate can be extracted as a minimal closed patterns which are also cliques in a graph constructed based on C1. For the computation, any algorithm for enumerating maximal cliques (e.g. [13]) can be used.

Identifying Candidate Pairs of Base Patterns: Then we identify the candidate pairs of base patterns in $Bases$ each of which satisfies the constraint C4. The set of candidates is denoted by $Pairs$.

Finding Chance Patterns for Each Pair of Base Patterns: For each candidate pair $(B_L, B_R) \in Pairs$, we try to find patterns which can be strongly supported by B_L and B_R. Such a pattern can be also extracted as a clique in another undirected graph constructed based on the constraint C5.

For more detailed discussion on each phase, readers can refer to [31].

4.6 Experimental Result

A system based on our framework for finding Top-N chance patterns has been implemented in C and compiled with gcc-4.4.5. Our experimentation has been carried out on a PC with Intel® Core™2 Duo processor (L7700:1.80 GHz) and 2 GB main memory.

Our dataset is a collection of articles appeared in a Japanese Newspaper "*Mainichi*" in 1994. Especially, it consists of articles in the category of "Economy". The number of articles (documents) is 9,810. After a standard preprocess, we have extracted 1,415-nouns as vocabulary words.

Under the setting [2], $max\_correl = 0.03$ and $min\_sim = 0.4$, we have tried to extract Top-3 chance patterns. One of the obtained chance patterns is as follows:

Chance Pattern: {`Business`, `Special`, `Improvement`} (Frequency : 2)

Base Pattern 1: {`Business`, `Release`, `Picture`, `Special`, `Compact`, `Equipment`, `Recreation`, `RV`, `Multipurpose`} (Frequency : 3)

Base Pattern 2: {`Price`, `Business`, `Release`, `Tokyo`, `Standard`, `Limited`, `Equipment`, `Nagoya`} (Frequency : 4)

Each document with the chance pattern is about a release information for a limited edition compact car with some particular improvement. The number of them (i.e. the frequency of the chance pattern) is 2. On the other hand, each document with Base Pattern 1 is about a release information for a multipurpose

[2] The parameter values have been determined based on our preliminary experimentations.

compact RV with a picture and that with Base Pattern 2 about a release information for a limited edition car sold in some limited areas including Tokyo and Nagoya. Their numbers are 3 and 4, respectively. The supporting power is 0.25.

Thus, our chance pattern is rare and shows some relationship between more frequent base patterns. However, the documents with Base Pattern 1 seem to be conceptually close to those with Base Pattern 2 while they are extensionally disjoint. Therefore, interestingness of the chance pattern would not be so high. We are required to introduce some additional constraints on conceptual difference between base patterns. Although we need to further improve our current method, the authors expect that our chance patterns have potential ability to reveal hidden implicit relationships among frequent patterns.

The computation time for finding Top-3 chance patterns with the above parameter setting is 125 seconds, including times for constructing graphs. We have first obtained 34 base patterns satisfying the constraints. From them, 55 pairs of base patterns have been extracted. Then, for each candidate pair, it has been checked whether chance patterns can be supported by the pair of base patterns with Top-N supporting powers. Although some pruning rules are available in our search, computational performance of our algorithm must be improved for much larger datasets. Particularly, we might need some additional constraints concerned with meaningfulness of patterns.

5 Concluding Remarks

In this paper, we overviewed the progress of our investigations on indicative concepts (closed patterns) and chance patterns. An indicative concept is defined as a concept with the weakly-correlated intent consisting of general attributes. Moreover, the intent has sufficient objective evidence.

In order to reveal a hidden relationship between a pair of more frequent patterns, we have extended the notion of indicative concepts by imposing some structural constraints on an indicative pattern and larger patterns bridged by it. Our pattern to be extracted is called a chance pattern and defined as an indicative pattern C which is supported by a pair of more frequent base patterns B_L and B_R. Thus, the chance pattern C is expected to imply or reveal some hidden relationship between B_L and B_R. From our preliminary experimentation, we have got a feeling that our chance pattern C would have potential ability of suggesting hidden relationships among base patterns B_L and B_R. In order to make our chance patterns more valuable, we can develop the current framework from several points of views.

If a chance pattern suggests a relationship between base patterns conceptually different, its interestingness and unexpectedness will be gained. It would be, therefore, worth investigating conceptual difference between base patterns.

In the current framework, we only focus on co-occurrence of a chance pattern and a base pattern as a supporting power. It might be required to take a *causal relationship* between a chance pattern and a base pattern into account so that our chance pattern can actually work as a valuable trigger for chance discovery. It is quite important to progress our future work in this direction.

Furthermore, we need to improve efficiency of our computation algorithms for larger scale datasets.

With further progress in these directions, the authors hope to make substantial contributions to the field of chance discovery.

References

1. Han, J., Cheng, H., Xin, D., Yan, X.: Frequent Pattern Mining - Current Status and Future Directions. Data Mining and Knowledge Discovery 15(1), 55–86 (2007)
2. Uno, T., Kiyomi, M., Arimura, H.: LCM ver. 2: Efficient Mining Algorithm for Frequent/Closed/Maximal Itemsets. In: Proc. of IEEE ICDM 2004 Workshop, FIMI 2004 (2004), `http://sunsite.informatik.rwth-aachen.de/Publications/CEUR-WS//Vol-126/`
3. Agrawal, R., Srikant, R.: Fast Algorithms for Mining Association Rules. In: Proc. of the 20th Int'l Conf. on Very Large Databases, VLDB 1994, pp. 487–499 (1994)
4. Wang, J., Han, J., Pei, J.: CLOSET+: Searching for the Best Strategies for Mining Frequent Closed Itemsets. In: Proc. of the 9th ACM SIGKDD Int'l Conf. on Knowledge Discovery and Data Mining, KDD 2003, pp. 236–245 (2003)
5. Gan, G., Ma, C., Wu, J.:Data Clustering – Theory, Algorithms, and Applications. SIAM (2007)
6. Szathmary, L., Napoli, A., Valtchev, P.: Towards Rare Itemset Mining. In: Proc. of the 19th IEEE Int'l Conf. on Tools with Artificial Intelligence, ICTAI 2007, pp. 305–312 (2007)
7. Troiano, L., Scibelli, G., Birtolo, C.: A Fast Algorithm for Mining Rare Itemsets. In: Proc. of the 2009 9th Int'l Conf. on Intelligent Systems Design and Applications, ISDA 2009, pp. 1149–1155 (2009)
8. Weiss, G.M.: Mining with Rarity: A Unifying Framework. SIGKDD Explorations 6(1), 7–19 (2004)
9. Ohsawa, Y.: Discovery of Chances Underlying Real Data. In: Arikawa, S., Shinohara, A. (eds.) Progress in Discovery Science. LNCS (LNAI), vol. 2281, pp. 168–177. Springer, Heidelberg (2002)
10. Ohsawa, Y., Benson, N.E., Yachida, M.: KeyGraph: Automatic Indexing by Segmenting and Unifing Co-occurrence Graphs. The IEICE Transactions on Information and Systems (Japanese Edition) J82-D-I(2), 391–400 (1999)
11. Maeno, Y., Ohsawa, Y.: Human-Computer Interactive Annealing for Discovering Invisible Dark Events. IEEE Transactions on Industrial Electronics 54(2), 1184–1192 (2007)
12. Ohsawa, Y., Yachida, M.: Discover Risky Active Faults by Indexing an Earthquake Sequence. In: Arikawa, S., Nakata, I. (eds.) DS 1999. LNCS (LNAI), vol. 1721, pp. 208–219. Springer, Heidelberg (1999)
13. Tomita, E., Akutsu, T., Matsunaga, T.: Efficient Algorithms for Finding Maximum and Maximal Cliques: Effective Tools for Bioinformatics. In: Biomedical Engineering. Trends in Electronics, Communications and Software, pp. 625–640. InTech (2011)
14. Newman, M.E.J.: Finding Community Structure in Networks Using the Eigenvectors of Matrices. Physical Review, E 74, 036104 (2006)
15. Barber, M.J., Clark, J.W.: Detecting Network Communities by Propagating Labels under Constraints. Physical Review, E 80, 026129 (2009)
16. Omiecinski, E.R.: Alternative Interest Measures for Mining Associations in Databases. IEEE Transactions on Knowledge and Data Engineering 15(1), 57–69 (2003)

17. Bay, S.D., Pazzani, M.J.: Detecting Group Differences: Mining Contrast Sets. Data Mining and Knowledge Discovery 5(3), 213–246 (2001)
18. Dong, G., Li, J.: Efficient Mining of Emerging Patterns: Discovering Trends and Differences. In: Proc. of the 5th ACM SIGKDD Int'l Conf. on Knowledge Discovery and Data Mining, KDD 1999, pp. 43–52 (2000)
19. Alhammady, H., Ramamohanarao, K.: Using Emerging Patterns and Decision Trees in Rare-Class Classification. In: Proc. of the 4th IEEE Int'l Conf. on Data Mining, ICDM 2004, pp. 315–318 (2004)
20. Novak, P.K., Lavrac, N.: Supervised Descriptive Rule Discovery: A Unifying Survey of Contrast Set, Emerging Pattern and Subgroup Mining. The Journal of Machine Learning Research Archive 10, 377–403 (2009)
21. Geng, L., Hamilton, H.J.: Interestingness Measures for Data Mining: A Survey. ACM Computing Surveys 38(3), Article 9 (2006)
22. Brin, S., Motwani, R., Silverstein, C.: Beyond Market Basket: Generalizing Association Rules to Correlations. In: Proc. of the ACM Int'l Conf. on Management of Data, SIGMOD 1997, pp. 265–276 (1997)
23. Ganter, B., Wille, R.: Formal Concept Analysis - Mathematical Foundations, 284 pages. Springer (1999)
24. Pasquier, N., Bastide, Y., Taouil, R., Lakhal, L.: Efficient Mining of Association Rules Using Closed Itemset Lattices. Information Systems 24(1), 25–46 (1999)
25. Nakajima, T.: Finding Concise Rare Concepts by Excavation of Pattern Pools Based on Local Branch-and-Bound Searches, Master Thesis, Graduate School of Information Science and Technology, Hokkaido University (2010) (in Japanese)
26. Haraguchi, M., Okubo, Y.: Pinpoint Clustering of Web Pages and Mining Implicit Crossover Concepts. In: Web Intelligence and Intelligent Agents, pp. 391–410. InTech (2010)
27. Li, A., Haraguchi, M., Okubo, Y.: Implicit Groups of Web Pages as Constrained Top-N Concepts. In: Proc. of the 2008 IEEE/WIC/ACM Int'l Conf. on Web Intelligence and Intelligent Agent Technology Workshops, pp. 190–194 (2008)
28. Taniguchi, T., Haraguchi, M.: Discovery of Hidden Correlations in a Local Transaction Database Based on Differences of Correlations. Engineering Application of Artificial Intelligence 19(4), 419–428 (2006)
29. Li, A., Haraguchi, M., Okubo, Y.: Contrasting Correlations by an Efficient Double-Clique Condition. In: Perner, P. (ed.) MLDM 2011. LNCS(LNAI), vol. 6871, pp. 469–483. Springer, Heidelberg (2011)
30. Okubo, Y., Haraguchi, M., Nakajima, T.: Finding Rare Patterns with Weak Correlation Constraint. In: Proceedings of the 2010 IEEE International Conference on Data Mining Workshops, ICDMW 2010, pp. 822–829 (2010)
31. Okubo, Y., Haraguchi, M., Hirokawa, S.: Finding Top-N Chance Patterns with KeyGraph®-Based Importance. In: König, A., Dengel, A., Hinkelmann, K., Kise, K., Howlett, R.J., Jain, L.C. (eds.) KES 2011, Part II. LNCS(LNAI), vol. 6882, pp. 457–468. Springer, Heidelberg (2011)
32. Okubo, Y., Haraguchi, M.: An Algorithm for Finding Indicative Concepts Connecting Larger Concepts Based on Structural Constraints. In: Contributions to ICFCA 2011, The 9th Int'l Conf. on Formal Concept Analysis, ICFCA 2011, pp. 53–68 (2011)
33. Okubo, Y., Haraguchi, M.: An Algorithm for Extracting Rare Concepts with Concise Intents. In: Kwuida, L., Sertkaya, B. (eds.) ICFCA 2010. LNCS(LNAI), vol. 5986, pp. 145–160. Springer, Heidelberg (2010)

Author Index

Abe, Akinori 1
Abe, Hidenao 215

Bardone, Emanuele 19

Furuhata, Masabumi 127

Gams, Matjaž 163

Hamaoka, Shuhei 63
Haraguchi, Makoto 231
Hong, Chao-Fu 91

Leung, Clement H.C. 143
Lin, Mu-Hua 91

Magnani, Lorenzo 19
Marsh, Ronald 191
Miwa, Kazuhisa 33
Mizuta, Takanobu 127

Nakajima, Takeshi 231
Nitta, Katsumi 79

Ogaard, Kirk 191
Ohsawa, Yukio 107
Okubo, Yoshiaki 231
Okuda, Kiyoshi 63

Pang, Chaoyi 143
Pogorelc, Bogdan 163

Rybakov, Vladimir 177

Sarlin, Peter 49
So, Jihei 127
Sunayama, Wataru 63

Tang, Xijin 143
Terai, Hitoshi 33
Tsumoto, Shusaku 215

Wang, Ai-Ling 91
Wang, Hao 107

Yang, Hsiao-Fang 91

Zhang, Hao Lan 143

Made in the USA
Las Vegas, NV
16 September 2026

653c387e-4acd-43c9-94f2-c2ffd0a719fdR02